MARINE BIOLOGY

TELEOSTS

PHYSIOLOGY, EVOLUTION AND CLASSIFICATION

MARINE BIOLOGY

Additional books in this series can be found on Nova's website under the Series tab.

Additional e-books in this series can be found on Nova's website under the e-Books tab.

MARINE BIOLOGY

TELEOSTS

PHYSIOLOGY, EVOLUTION AND CLASSIFICATION

MIKAEL HERLEIF
EDITOR

nova
science publishers
New York

Library of Congress Cataloging-in-Publication Data

ISBN: 978-1-53613-660-9

Published by Nova Science Publishers, Inc. † New York

CONTENTS

Preface

The authors begin this compilation by analyzing catalase and peroxidase enzymes in different Black Sea teleosts related to their taxonomic, physiological and ecological position and evaluation of the anthropogenic impact on these antioxidant enzymes in fish tissues. The complex of specific phylogenic, physiological and ecological features of fish species may modify their antioxidant status, and it is important to understand for the development of monitoring programs. Next, the ratio of constitutive and reparative neurogenesis in the pallium of the juvenile masu salmon, *Oncorhynchus masou*, is analyzed. Since salmonid juveniles are characterized by a high level of constitutive neurogenesis, the authors investigated proliferative activity and neurodifferentiation in the dorsal pallial area of the masu salmon juvenile. The study has allowed for identification of the superficially located periventricular proliferative zone, which corresponds to the pallial periventricular zone of other fish species. The objective of the final chapter was to study proliferation processes and the role of radial glia and neural stem cells in the event of injurious action on cerebellum of masu salmon juveniles. Using the immunoperoxidase staining of the glial fibrillary acidic protein, doublecortin and proliferating cells nuclear antigen, processes of proliferation and gliogenesis after mechanical trauma of cerebellum of masu salmon juveniles were studied.

Chapter 1 - Metabolism of the living organisms is associated with oxygen consumption. Microbes, plants and animals use oxygen in the major physiological and biochemical pathways. However, O_2 has two unpaired electrons and the univalent reduction of the molecular oxygen produces reactive oxygen species (ROS), which lead the oxidative stress and damage biomolecules, such as lipids, proteins, DNA, cells and tissues. Living organisms have antioxidant defense to counteract the toxic effects of ROS, and it plays a key role in inactivation of ROS, and thereby controls of oxidative stress as well as redox signaling. Among antioxidant enzymes catalase (CAT) (EC 1.11.1.6.) which catalyzes the reduction of hydrogen peroxide to water and total peroxidase (PER) (the sum of SeGPx (EC 1.11.1.9) and selenium-independent GPx (EC 1.11.1.9) activities), which reduces both hydrogen peroxide and organic hydroperoxides are the main antioxidant enzymes. The study of these enzymes in different taxonomic groups of the living organisms is very important for the understanding of the mechanisms of the evolutionary process and adaptations of the species for environmental conditions. Antioxidant status of the fish depends on many biotic and abiotic factors, such as temperature, season, diet, salinity, oxygen concentration and many others, as well as on their taxonomic position and life cycle. It reflects the adaptive strategy of fish species to oxidative stress and their ability to cope with the environment. At the other side, antioxidant status is attributed with phylogenetic position and ecological peculiarities of the organism, and these parameters in primitive life are differed compared to birds and mammals. The levels of antioxidant enzymes activity are ranged in fish tissues, and they depend on the tissues specificity and physiological function. The interspecies variations of the enzymes may reflect the specific adaptations to the oxidative stress and the protective mechanisms against ROS damage in fish, belonging to different ecological groups and inhabiting different geographical locations. Studies of antioxidant enzymes in different species are very important for the understanding the response mechanisms against pollution in fish of various taxonomic and ecological groups as well as for aquaculture purposes, and for receiving the information of repair mechanisms and adaptation to unfavorable factors. The combination of specific phylogenetic, physiological and ecological features of fish species may modify antioxidant status, and it is important for development of monitoring programs. The aim of the present work is to analyze catalase and peroxidase enzymes in different Black Sea teleosts related to their taxonomic, physiological and ecological position and for the purpose of the use of these antioxidant enzymes in fish tissues as biomarkers for the anthropogenic impact on marine environment.

Chapter 2 - The chapter considers the ratio of constitutive and reparative neurogenesis in pallium of juvenile masu salmon *Oncorhynchus masou*. Since salmonid juveniles are characterized by a high level of constitutive neurogenesis, the authors investigated the proliferative activity and neurodifferentiation in the dorsal pallial area of juvenile masu salmon. The study of proliferative activity in pallium of juvenile masu salmon *O. masou* has allowed us to identify the superficially located periventricular proliferative zone, which corresponds to the pallial periventricular zone (PVZ) of other fish species, including the dorsal, lateral, and medial compartments. The PCNA+ cells are also identified in parenchyma of masu salmon intact brain, and their maximum concentration is observed in the medial zone. In intact brain, solitary proliferating cells were identified in parenchymal regions with their maximum concentration in the medial zone. After a mechanical injury, the zones of reparative neurogenesis — neuroepithelial neurogenic niches and zones of reactive neurogenesis surrounded by radial glial fibers—appear in masu salmon pallium. An increase in the proliferative activity is characteristic of all compartments of the PVZ (especially of the medial zone), including its deep parenchymal layers. The PVZ of juvenile masu salmon contains clusters of undifferentiated HuCD+ neurons. A change in the HuCD+ cell topography is observed in the mechanically injured masu salmon pallium, in particular, reactive neurogenic niches in the lateral zone and an increase in the cell distribution density and cell migration patterns in the medial zone. A high level of constitutive neurogenesis is characteristic of juvenile masu salmon brain. Undifferentiated HuCD+ neurons form clusters in the PVZ of intact juvenile masu salmon brain. Neurons at different stages of maturation, differing in the level of HuCD activity, were detected in the parenchymal layers of the dorsal region. A mechanical injury of masu salmon pallium is accompanied by a change in the topography of HuCD+ cells. Formation of reactive neurogenic niches is characteristic of the lateral zone, while the medial zone displays an increase in the density of immunopositive and negative cells, as well as a distinct pattern of cell migration. HuCD and PCNA immunolabeling patterns in intact pallium suggest intensive constitutive neurogenesis in the pallial PVZ of juvenile masu salmon. After a mechanical injury, the neurogenic activity spreads to deeper layers of parenchyma. The main sources of new neurons during the reparative neurogenesis are reactive neurogenic niches.

Chapter 3 - The objective of this chapter was to study proliferation processes and the role of radial glia and neural stem cells in the event of injurious action on cerebellum of masu salmon *Oncorhynchus masou* juvenile. Using the immunoperoxidase staining of the glial fibrillary acidic protein (GFAP), doublecortin (DC) and proliferating cells nuclear antigen (PCNA), processes of proliferation and gliogenesis after mechanical trauma of cerebellum of juvenile masu salmon O. masou were studied. After the trauma, the intensity of proliferation and migration processes varies in different zones. Proliferation processes decrease after the trauma in lateral and basal zones, and migration increases. In the dorsal zone, on the contrary, migration processes significantly decrease and proliferation increases. In the dorsal matrix zone of cerebellum, intense cell proliferation was detected. In the dorsal, lateral, and basal zone of the molecular layer of cerebellum after traumatic injury, neurogenic niches containing PCNA and cells, as well as a heterogeneous population of PCNA- cells, were identified. At the location of neurogenic niches, fibers of radial glia and small single, intensely or moderately labeled GFAP cells were discovered. As a result of injury, GFAP+ fibers of radial glia, which form differently directed, radially oriented bundles, appeared in the dorsal matrix zone. Such structural formations were not discovered in intact animals. The authors suppose that, after the trauma, a structural reconstruction, connected with partial spatial reorientation of the radial glia fibers, and formation of specific directions for cells formed in this zone occur in the dorsal matrix zone. As a result of the trauma, elements of the radial glia, including both the cells possessing typical morphology and the cell fragments represented by long radially oriented processes or cell body, containing initial fragments of radial fibers, appeared in masu salmon's cerebellum.

Chapter 4 - The retina of teleosts has become a powerful model system in which to study different aspects of development and regeneration. While retinal development is completed during embryogenesis in most classes of vertebrates, it continues in teleosts throughout much of the animal´s life. The mature retina grows from two specialized niches that sustain retinal stem cells (RSC): (i) the ciliary marginal zone (CMZ), a narrow ring of tissue located in the most-peripheral region of the retina that is composed of stem cells and progenitor cells that generate new neurons added to the periphery of the growing neural tissue, and (ii) the Müller glial cells, the radial glia of the neural retina, which generate neuronal progenitor cells that continue to proliferate and migrate to the photoreceptor layer, where they become precursor cells that are committed to differentiating into rod photoreceptors. Additionally, in the teleost retina, Müller glia have neurogenic capacity and mediates regeneration when stimulated by acute damage, by re-entering the cell cycle and re-establishing retinal cell types.

It has been suggested that the regenerative response of retinal tissue to injuries diminishes through the course of evolution from fish to mammals. Thus, the addition of new neurons from RSC is diminished in chickens compared to that observed in fish and amphibians. In contrast, the neuro-regenerative capacity of RSC in mammals is minimal or non-existent and therefore, degenerative retinal diseases in humans are irreversible. There are many genes, including structural proteins, signaling pathways and transcription factors that are expressed by normal RSC in teleosts that may be related to maintaining progenitor-like properties. Understanding the combinatorial expression of genes involved in the stimulation of the formation of neurogenic RSC in teleostean experimental models might lead to new genetic treatments for retinal degenerative diseases in humans.

In: Teleosts
Editor: Mikael Herleif

ISBN: 978-1-53613-660-9

Chapter 1

Catalase and Peroxidase in Black Sea Teleosts: Evolutionary, Systematical, and Physiological Peculiarities

Irina I. Rudneva*, Irina I. Chesnokova and Tatiana B. Kovyrshina
The A. O. Kovalevsky Institute
of Marine Biological Research of RAS,
Sevastopol, Russia

Abstract

Metabolism of the living organisms is associated with oxygen consumption. Microbes, plants and animals use oxygen in the major physiological and biochemical pathways. However, O_2 has two unpaired electrons and the univalent reduction of the molecular oxygen produces reactive oxygen species (ROS), which lead the oxidative stress and damage biomolecules, such as lipids, proteins, DNA, cells and tissues. Living

* Corresponding Author Email: svg-41@mail.ru.

organisms have antioxidant defense to counteract the toxic effects of ROS, and it plays a key role in inactivation of ROS, and thereby controls of oxidative stress as well as redox signaling. Among antioxidant enzymes catalase (CAT) (EC 1.11.1.6.) which catalyzes the reduction of hydrogen peroxide to water and total peroxidase (PER) (the sum of SeGPx (EC 1.11.1.9) and selenium-independent GPx (EC 1.11.1.9) activities), which reduces both hydrogen peroxide and organic hydroperoxides are the main antioxidant enzymes. The study of these enzymes in different taxonomic groups of the living organisms is very important for the understanding of the mechanisms of the evolutionary process and adaptations of the species for environmental conditions. Antioxidant status of the fish depends on many biotic and abiotic factors, such as temperature, season, diet, salinity, oxygen concentration and many others, as well as on their taxonomic position and life cycle. It reflects the adaptive strategy of fish species to oxidative stress and their ability to cope with the environment. At the other side, antioxidant status is attributed with phylogenetic position and ecological peculiarities of the organism, and these parameters in primitive life are differed compared to birds and mammals. The levels of antioxidant enzymes activity are ranged in fish tissues, and they depend on the tissues specificity and physiological function. The interspecies variations of the enzymes may reflect the specific adaptations to the oxidative stress and the protective mechanisms against ROS damage in fish, belonging to different ecological groups and inhabiting different geographical locations. Studies of antioxidant enzymes in different species are very important for the understanding the response mechanisms against pollution in fish of various taxonomic and ecological groups as well as for aquaculture purposes, and for receiving the information of repair mechanisms and adaptation to unfavorable factors. The combination of specific phylogenetic, physiological and ecological features of fish species may modify antioxidant status, and it is important for development of monitoring programs. The aim of the present work is to analyze catalase and peroxidase enzymes in different Black Sea teleosts related to their taxonomic, physiological and ecological position and for the purpose of the use of these antioxidant enzymes in fish tissues as biomarkers for the anthropogenic impact on marine environment.

1. Introduction

The introduction of molecular oxygen into Earth's atmosphere billions of years ago has allowed animals to produce energy more efficiently (Falkowski et al., 2004). Metabolism of the living organisms is associated with oxygen consumption. Microbes, plants and animals use oxygen in the major physiological and biochemical pathways. Living organisms obtain and use oxygen, and evolved over hundreds or millions of years together with the evolution of oxygen-delivery systems. A steady supply of oxygen to the existence of organism was very important, because its metabolism depends on oxygen as a primary source of fuel (Fisher & Burggren, 2007). However, O_2 has two unpaired electrons and the univalent reduction of the molecular oxygen produces reactive oxygen species (ROS), which lead the oxidative stress and damage biomolecules, such as lipids, proteins, DNA, cells and tissues. Living organisms have antioxidant defense to counteract the toxic effects of ROS, and it plays a key role in inactivation of ROS, and thereby controls of oxidative stress as well as redox signaling. A number of mechanisms have evolved to defend against oxygen deprivation (hypoxia and anoxia) and for the hyperoxic conditions. At the other side, ROS can play an essential role in the evolution of living organisms, because the growth of oxygen level in the atmosphere in the combination with the increased mobility and food intake, led to increase cellular level of free radicals, ROS interacted with DNA of the primitive life, enhancing mutations and providing new regulatory mechanisms and biodiversity (Yang et al., 2018).

Recent years the role of oxidative stress in the life history has been received much attention, caused the effects of the climate changes and anthropogenic pollution, and their consequences for environment and biota. However, generally the researchers focus their studies on high taxonomic groups, while the information about fish and invertebrates is less (Birnie-Gauvin et al., 2017).

Antioxidant system presents in all living organisms and it includes low molecular weight scavengers (vitamins A, E, K and C, carotenoids, SH-containing substances (glutathione) and small-molecule antioxidants namely uric acid, urea, and etc.), and special adopted enzymes (catalase, superoxide dismutase, peroxidases, especially selenium-dependent glutathione peroxidase (SeGPx) (EC 1.11.1.9), glutathione reductase (GR) (EC 1.8.1.7) and some others) (Winston, 1991). Heme peroxidases and catalases are the key enzymes of hydrogen peroxide metabolism and signaling, which present in all kingdoms of life (Zamocky et al., 2014). Among them catalase (CAT) (EC 1.11.1.6.) and total peroxidase (PER) (the sum of SeGPX and selenium-independent glutathione peroxidase (GPx) (EC 1.11.1.9) activities), are the key antioxidant enzymes.

CAT catalyzes the reduction of hydrogen peroxide to water:

$$2H_2O_2 \rightarrow 2H_2O + O_2$$

Heme peroxidases are the most abundant family of peroxidase catalyzing the H_2O_2-dependent oxidation of a wide kinds of substrates. They involved in innate immune response, hormone and prostaglandin synthesis crosslinking of protein extracellular matrix (Øvergård et al., 2017). Additionally, they play a role in the process of the fish eggs fertilization. Total PER (SeGPx and GPx) reduces both hydrogen peroxide and organic hydroperoxides:

$$ROOH + 2GSH \rightarrow ROH + H_2O + GSSG \text{ (SeGPx)}$$
$$H_2O_2 + 2\,GSH \rightarrow 2\,H_2O + GSSG$$

GPx eliminates the peroxides produced as effect of reactions of biomolecules with ROS. There are several types of GPx, which was isolated and cloned in various organisms (Liu et al., 2018).

Therefore, study of these enzymes and their activity in different taxonomic groups of the living organisms is very important for the understanding of the mechanisms of the evolutionary process and adaptations of the species to environmental conditions. Fishes are highly distributed in various water bodies with great variability of their conditions, including salinity, oxygen concentration, temperature, and etc. (Birnie-Gauvin et al., 2017). They play an important role in the aquatic ecosystems and for people, because they are the main source of seafood, which has been recognized as a high-quality, healthy and safe food type. Fish tissues are rich of the antioxidants, and some of them are present in seafood in high concentrations. Therefore, they are the main sources of these essential components for people diet and one of the most important food commodities consumed worldwide. At the other side, antioxidant enzymes namely catalase and peroxidase are good biomarkers and they are used as a tool for the evaluation of stress conditions in the wild and aquaculture fish populations (Mukherjee et al., 2017; Alti et al., 2016).

Antioxidant status of fish depends on many biotic (food consumption, reproductive strategy, aging and senescence, belonging to different ecological groups), and abiotic factors such as temperature, season, salinity, oxygen concentration in water, and many others as well as on their taxonomic position and life history. It reflects the adaptive strategy of fish species to oxidative stress and their ability to cope with the environment. At the other side, antioxidant status is associated with phylogenetic position and ecological peculiarities of the organisms, and these parameters in primitive life are differed compared to birds and mammals. The levels of the antioxidant enzymes are ranged in fish tissues, and they depend on the tissue specificity and physiological function. The interspecies variations of the enzymes may reflect the specific adaptations to the oxidative stress and protective mechanisms against ROS damage in fish, belonging to different ecological groups and inhabiting different geographical locations and biotops. Thus, the comparative studies of antioxidant defense in teleosts are very important for the understanding the main ways of evolutionary process and the mechanisms of adaptation to aerobic conditions (Filho & Boveris, 1993; Filho et al., 1993; Filho, 1996; Sole et al., 2009).

Study of the antioxidant enzymes in different species is very important for the understanding the defense mechanisms against pollution in fish of various taxonomic and ecological groups, as well as for aquaculture purposes, and for receiving the information of repair mechanisms and adaptation to unfavorable impact. Combination of specific phylogenetic, physiological and ecological features of fish species may modify antioxidant status, and it is important for the creation of monitoring programs. The aim of the present work is to analyze the key antioxidant enzymes catalase and peroxidase in different Black Sea teleosts, belonging to different taxonomic groups related to their phylogenetic position, physiological, developmental and ecological peculiarities and anthropogenic impact.

2. Enzyme Activities and Fish Phylogenic Position

Tested Black Sea fish species were grouped according their phylogenetic position (Table 1).

Table 1. Phylogenic position of tested Black Sea teleosts (the classification is present according Boltachev and Karpova, 2017)

Order	Family	Fish species
Gadiformes	*Gadidae*	*Merlangius merlangus euxinus* (Nordmann, 1840)
	Lotidae	*Gaidropsarus mediterraneus* (Linnaeus, 1785)
Ophidiiformes	*Ophidiidae*	*Ophidion rochei* Muller, 1845
Mugiliformes	*Mugillidae*	*Liza aurata* (Risso, 1810)
Perciformes	*Sciaenidae*	*Sciaena umbra* Linnaeus, 1785
Perciformes	*Sparidae*	*Diplodus annularis* (Linnaeus, 1785)
Perciformes	*Centracanthidae*	*Spicara flexuosa* Rafinesque, 1810
Perciformes	*Labridae*	*Symphodus tinca* (Linnaeus, 1785)
Perciformes	*Carangidae*	*Trachurus mediterraneus ponticus* Aleev, 1956
Perciformes	*Gobiidae*	*Neogobius melanostomus* (Pallas, 1814) *Mesogobius batrachocephalus* (Pallas, 1814)
Perciformes	*Mullidae*	*Mullus barbatus ponticus* Essipov, 1927
Perciformes	*Uranoscopidae*	*Uranoscopus scaber* Linnaeus, 1785
Scorpaenoformes	*Scorpaenodae*	*Scorpaena porcus* Linnaeus, 1785
Pleuronectiformes	*Pleuronectidae*	*Platichthys luscus* (Pallas, 1814)

The most ancient orders are recognized as *Gadiformes* and *Ophidiiformes*, the following is *Mugiliformes,* then *Perciformes,* the orders *Scorpaenoformes* and *Pleuronectiformes* are relatively recent taxonomic groups. We studied CAT and PER activity in fish red blood cells (RBC) and in the liver related to their phylogenetic position. CAT activity in the RBC of tested Black Sea teleosts tended to increase in the species, belonging to the order *Perciformes* and *Scorpaenoformes* as compared to the species of ancient orders namely *Gadiformes* and *Ophidiiformes* (Figure 1). At the other side, CAT activity in *Platichthys luscus,* belonging to very specific order *Pleuronectiformes* was lower as compared with the species of *Perciformes.* Additionally, we could note that enzyme level varied significantly among the various species of *Perciformes,* which could depend on their biological and ecological peculiarities (see below).

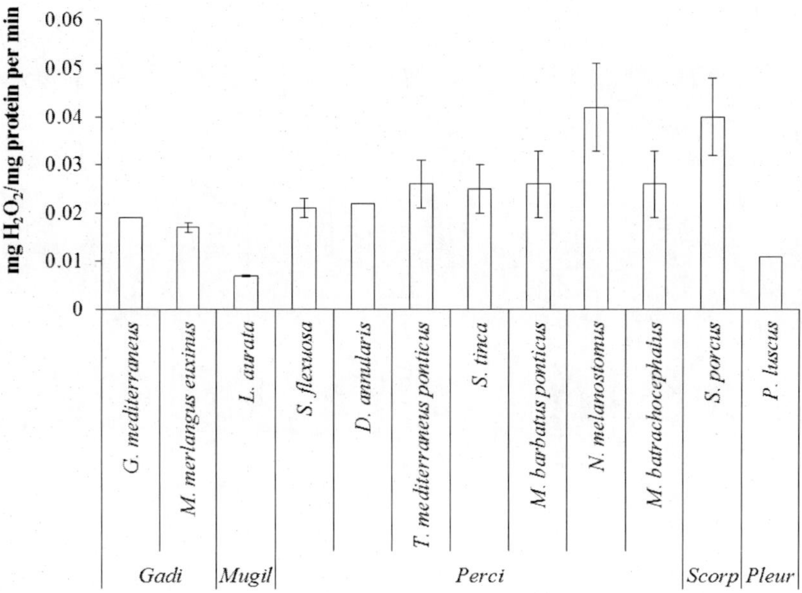

Figure 1. CAT activity in RBC of tested Black Sea fish species belonging to different orders: Gadi – *Gadiformes,* Mugil – *Mugiliformes*, Perci – *Perciformes*, Scorp – *Scorpaeniformes*, Pleur – *Pleuronectiformes*.

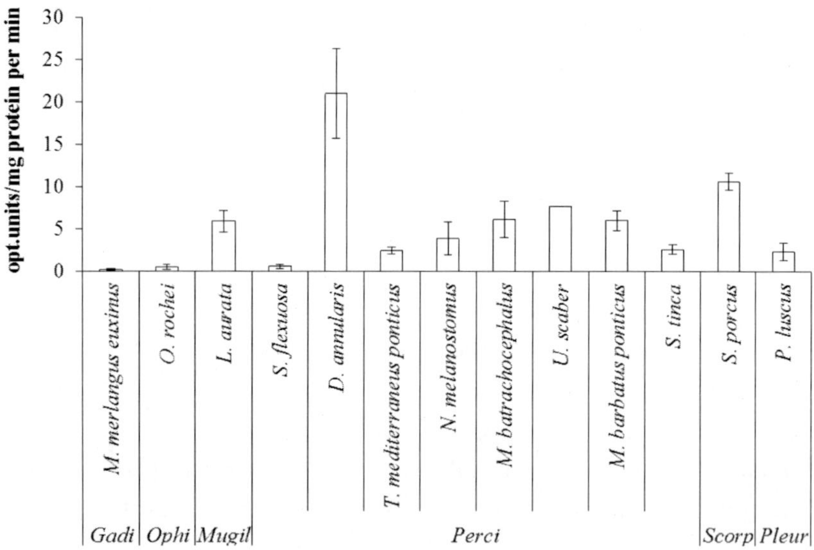

Figure 2. PER activity in RBC of tested Black Sea fish species belonging to different orders: Gadi – *Gadiformes,* Ophi – *Ophidiiformes,* Mugil – *Mugiliformes*, Perci – *Perciformes*, Scorp – S*corpaeniformes*, Pleur – *Pleuronectiformes*.

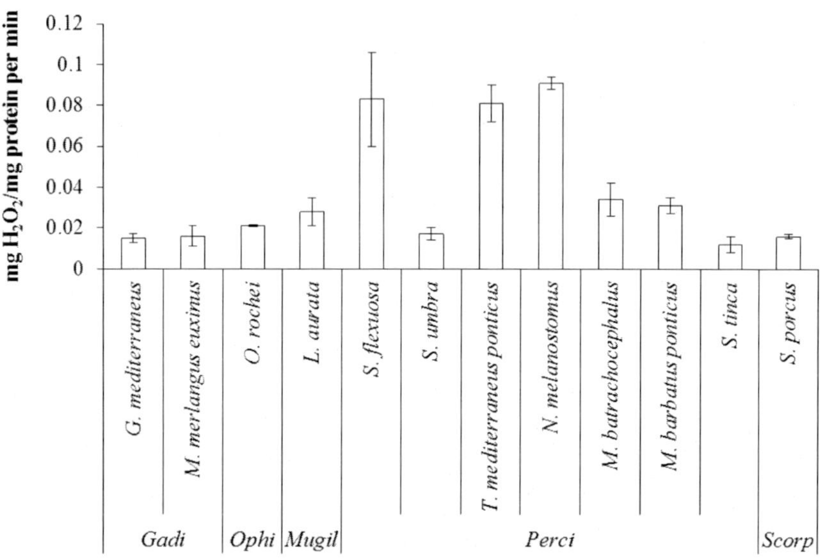

Figure 3. CAT activity in the liver of tested Black Sea fish species belonging to different orders: Gadi – *Gadiformes,* Ophi – *Ophidiiformes,* Mugil – *Mugiliformes*, Perci – *Perciformes*, Scorp – S*corpaeniformes*, Pleur – *Pleuronectiformes*.

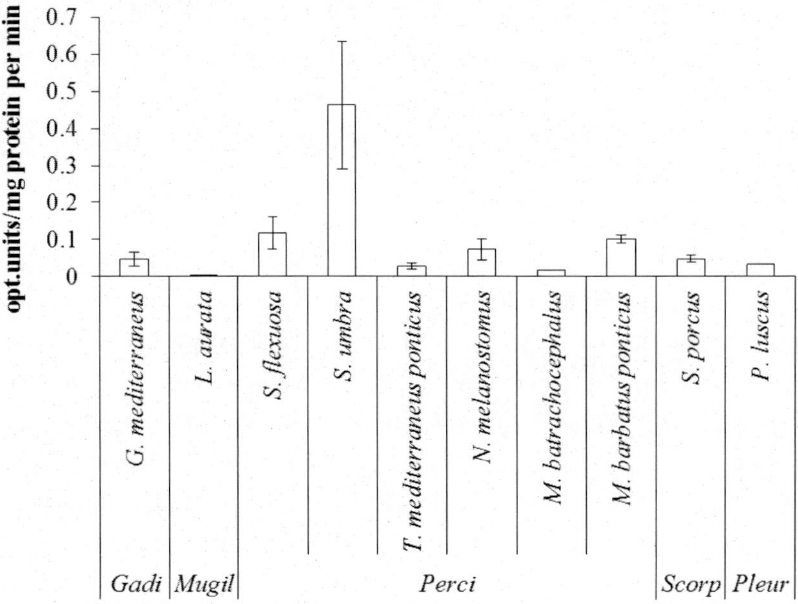

Figure 4. PER activity in the liver of tested Black Sea fish species belonging to different orders: Gadi – *Gadiformes,* Ophi – *Ophidiiformes,* Mugil – *Mugiliformes*, Perci – *Perciformes*, Scorp – *Scorpaeniformes*, Pleur – *Pleuronectiformes*.

The lowest PER activity was indicated in the RBC of examined fish species belonging to *Gadiformes* and *Ophidiiformes*, in the other tested teleosts enzyme activity varied significantly, especially in the animals of the order *Perciformes* (Figure 2).

In the liver high CAT activity was observed in the *Perciformes* species (Figure 3), and it varied widely. Enzymatic activity in the RBC of the species belonging to the other examined orders was lower, however, it was comparable with the *Perciformes* fishes. PER activity in the liver ranged widely and we could not find any phylogenetic-related trends (Figure 4).

In our previous studies we showed the lack of CAT activity in the RBC in elasmobranch *Squalus acanthias* related to several examined Black Sea teleosts. We proposed that low molecular weight antioxidants, especially SH-containing substances play an important role in the antioxidant defense in elasmobranch, which compensate the lack of CAT and low activity of superoxide dismutase (SOD) and glutathione reductase (GR). As we documented previously, enzymatic activity was less in elasmobranch tissues as compared with the teleost species, excepted PER activity in muscle, which was the greatest among the all examined fish (Rudneva, 1997a; 2012).

A comparative study of hepatic antioxidant defense enzymes in teleosts and elasmobranchs demonstrated, that their content in primitive fish species was lower than that in teleosts, and seems to follow the overall metabolic oxygen consumption or activity level from each fish major taxonomic group (Filho & Boveris, 1993; Filho et al., 1993). Decrease of antioxidant enzymes level and high content of low molecular weight antioxidants in elasmobranch tissues related to teleosts was reported by other investigators (Martinez-Alvarez et al., 2005; Sole et al., 2009). Therefore, our findings provide that more primitive species such as elasmobranch and species, belonging to ancient teleost orders have more primitive antioxidant system in which non-enzymatic compounds play an important role in defense organism against oxidative stress.

The differences between CAT and PER activities can be linked with the differences of the enzymes functions: PER is non-specific enzyme as compared with CAT, and it reduces both H_2O_2 and organic peroxides (see section *Introduction*), and it could compensate the lack of CAT concentration. In general, our findings demonstrated that antioxidant enzyme levels in tested Black Sea teleost species depending on their taxonomic position, however, despite this, we could suggest that biological and ecological peculiarities of fish species may play more important role in antioxidant response than phylogenetic ones.

3. Biological Peculiarities of Fish

3.1. Early Development and Ageing

Antioxidant enzyme activity was found in developing eggs of the fish. Ovoperoxidase was detected in the eggs, and it uses hydrogen peroxide as substrate, which is generated NADPH-oxidase system on the surface of egg shell (Takahashi et al., 1989; Winston, 1991). The authors demonstrated the changes in antioxidant enzyme activities in various fish species during early life. However, the trends of CAT and PER activities in various fish were not uniform and depended on species and developmental stage. Oxygen availability is one of the factor effecting hatching. Hypoxia or agents that cause hypoxia (low sea water dissolved concentration of cyanide or high concentrations of dissolved hydrogen), stimulate hatching whereas hyperoxia may suppress hatching. Therefore, antioxidant enzymes level varied during fish embryogenesis and in hatching larvae.

In our studies we found, that during embryogenesis of several Black Sea gobies *Neogobius melanostomus*, *Proterorhinus marmoratus* and blenny *Blennius sanguinolentus,* majority of enzymes, including CAT and PER, tended to increase in developing eggs and especially in hatching larvae, while the levels of low molecular scavengers decreased (Rudneva, 1999; 2014). Our results agree with the data of other investigators, who reported the similar trends of CAT and PER levels in fish early development at the case of *Solea senegalensis* (Sole et al., 2004), *Scophthalmus maximus* (Peters & Livingstone, 1996), *Dentex dentex* (Mourente et al., 1999), *Paralichthys olivaceus* (Cao et al., 2010), *Lates calcarifer* (Kalaimani et al., 2008). Among the H_2O_2 -removing systems in *Salmo iridaeus* embryo development CAT seems to be the only enzyme presents at significant level and plays more important role in preventing H_2O_2 damage in fish embryo as compared with GPx. During further fish development, activity of antioxidant enzymes CAT and GPx increased with the maximum at stage 33, when the formation of the swim bladder as well as pigmentation has occurred. At the end of the development the activity of glutathione-containing enzymes increased progressively and values were found to be about 85-312 fold higher than in early stages, while the variations of CAT activity was lower (Aceto et al., 1994).

Therefore, we could conclude, that the increasing rate of oxygen uptake may elevate pro-oxidant processes such as ROS generation, and increase in CAT and SeGPx activities with development of embryos to larvae. They would indicate a progressive need to degrade H_2O_2 and peroxides from tissues (Peters & Livingstone, 1996). The increase levels of CAT and GPx may be in response to other metabolic changes occurring at this time, possibly leading to change sources of oxidative stress.

Further aging changes can be attributed to development, genetic defects, the environmental unfavorable factors and diseases. The accumulation of the negative changes and the loss of the ability to repair or detoxify them increase the risk of death (Harman, 2003). The study of aging strategy, both endogenous and exogenous factors, which it modulate in different species is very important for the understanding of the key mechanisms of aging and their early prevention. Several studies have shown that oxidative stress led the environmental factors is involved in age-related processes and modulate them. Oxygen stress relates to immune system dysfunction seems to have an important role in senescence, in agreement with the oxidation/inflammation theory of aging (De La Fuente et al., 2005).

There are many publications on the effect of age on antioxidant enzyme activities in laboratory animals including various mammalian and bird species, and perhaps for aquaculture fishes, while the information of wild species is very limited. However, the aging processes in wild fish populations are associated with many environmental factors, such as anthropogenic impact and the abilities to adapt to the changing living conditions, especially in recent decades, caused with climate changes. The life span of species might be influenced by its metabolic rate and its increase correlates with the high radical production and with oxidative damage (Zelinski & Portner, 2000). Whereas these aspects have been amply studied in mammals and especially in humans, information on fish and other aquatic animals is very limited.

Fish erythrocytes are an excellent model for the investigations of ROS and antioxidant status, especially for age-related studies, because in these cells mitochondria and nucleus were shown (Filho, 2007). In our previous studies (Rudneva et al., 2010a) we found, that the trends in antioxidant enzyme activities in blood of Black Sea teleosts were not uniform, and we showed three types of responses with aging increase, which depended on the enzyme specificity and fish species: 1. enzymatic activity increased with age; 2. enzymatic activity decreased with age and 3. enzymatic activity was not changed with age. For instance, we observed strong correlation between CAT activity and fish age in *M. merlangus euxinus* and *T. mediterraneus ponticus*, significant values was shown in *G. mediterraneus* and moderate in *M. barbatus ponticus*, and low in *S. flexuosa*. The highest negative correlation coefficients were shown between PER activity and age in *G. mediterraneus* and *T. mediterraneus ponticus*, in *S. porcus* it was significant and moderate in *M. merlangus euxinus*. CAT activity showed no age dependence in blood of *N. melanostomus*, *S. porcus* and *S. flexuosa*. Age changes in CAT level were not evident in rainbow trout *Oncorhynhus mykiss* and black bullhead *Ameiurus melas* also (Otto & Moon, 1996). *G. mediterraneus* and *T. mediterraneus ponticus* enzymatic activity decreased with the age, while in *M. merlangus euxinus* the opposite tendency was observed. In *M. barbatus ponticus* CAT activity elevated in the middle-age fish, then decreased progressively in the elder ones. The similar tendency was found in the blood cells of sturgeon *Acipenser naccarii* (Martinez-Alvarez et al., 2005). Our findings agree with the results of the other studies of age-related CAT activity in various animals. The investigators documented that CAT activity has been decreased (Semsei et al., 1989) or unchanged with the age in brain of albino rats (Yargicoglu et al., 1999). Age- and size-related changes of CAT activity was shown in mollusk *Perna viridis* (Lau & Wong, 2003).

High negative correlation was indicated between PER activity and fish age in the RBC of *G. mediterraneus* and *T. mediterraneus*, in *S. porcus* and *M. merlangus euxinus* the values were positive and moderate, while in the other examined species the correlations were not observed. In this case we could agree with the opinion of several authors who proposed, that CAT and PER activity in the brain and in the liver of freshwater murrel *Channa punctatus* were mostly maturation-related rather than senescence-related (Nayak et al., 1999, cited by Martinez-Alvares et al., 2005).

Therefore, the age alterations of antioxidant enzyme activities in fish blood are the similar as compared with the other animals including terrestrial species. It may be the result of detrimental effect of life span that is conclude with the opinion of other researchers (Zelinski & Portner, 2000; Ahmed, 2005).

Our recent studies showed that in RBC of the several Black Sea teleosts (*S. porcus* and *T. mediterraneus ponticus*) CAT and PER level did not vary uniform related to age (Figure 5). Low correlation was detected between age and CAT activity in the blood of scorpion fish (r = 0.32), correlation between PER and fish age was greater (r = 0.62). We noted high age-related relationships between CAT and PER level in RBC of *S. porcus* (r = -0.73). Therefore, activity progressively decreased in the RBC of scorpion fish *S. porcus,* while in horse mackerel *T. mediterraneus ponticus* the values did not change. At the other side, the ratio of CAT/PER activity ranged between 0.002-0.007 in RBC of scorpion fish and 0.009-0.008 in horse mackerel.

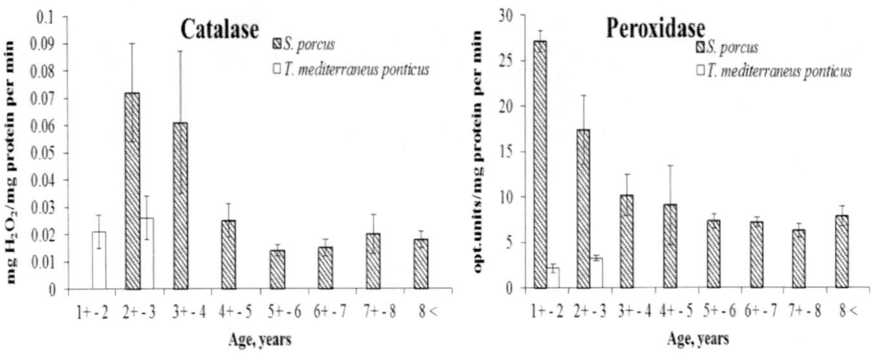

Figure 5. Age-related CAT and PER activity in RBC of Black Sea teleosts *S.porcus* and *T. mediterraneus ponticus*.

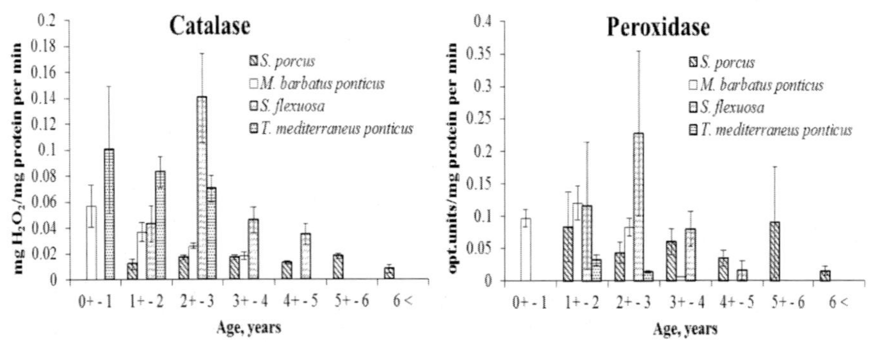

Figure 6. Age-related CAT and PER activity in the liver of Black Sea teleosts *S. porcus*, *M. barbatus ponticus, S. flexuosa*, *T. mediterraneus ponticus.*

In the liver of tested Black Sea teleosts CAT and PER activity varied related to age, tended to decrease in the major of tested fish species (Figure 6).

Age-related correlation of CAT activity was estimated as r = 0.31 in the liver of *S. porcus,* in red mullet *M. barbatus pointicus* the corresponding value was higher r = 0.93. No age-related correlations we found in hepatic CAT activity in high body pickerel *S. flexuosa* and horse mackerel *T. mediterraneus ponticus.* PER activity in the liver of tested teleosts demonstrated low age-related correlation (r = 0.17), at the case of scorpion fish and high body pickerel, in the liver of red mullet *M. barbatus ponticus* it was higher (r = 0.50). Great age-related correlation (r = 0.75 and r = 0.90) was observed between CAT and PER activity in *S. porcus* and *S. flexuosa* correspondingly.

These findings agree with the data of other investigators, who described the tendency of decrease of enzymatic activity in the erythrocytes and in the liver of old fish as compared with younger ones (Otto & Moon, 1996). It was linked with accumulation of oxidative products in organism and decrease of its defense abilities. At the other side, the antioxidant system in long-lived fish species has evolved to increase its capacity with age, while the other species have constant enzyme activities, or they decline throughout their life. These differences could play important role in the regulation of the mechanisms of the lifespan of teleost fish species.

However, we could note the significant differences in antioxidant enzyme responses in examined fish species, which were attributed with the peculiarities of their ecological and physiological characteristics (see section 3.2). For instance, in our previous publication, we suggested that age alterations were much more significant in benthic and suprabenthic fish species, compared with suprabenthic/pelagic and pelagic ones. We could propose that benthic and suprabenthic forms live in more contaminant environment, because many pollutants accumulate in bottom sediments and low water layers and they impact aquatic organisms (Rudneva et al., 2010b).

3.2. Reproductive Strategy and Gender Peculiarities

Aerobic organisms, in the process of the evolution have adapted to use their living resources (e.g., energy, time, nutrients, etc.) for reproduction. They have evolved different strategy of breeding frequently, fecundity, egg hatching rate and larval survival. The costs associated with reproduction have been attributed with the increase of metabolic rate and high production in ROS (Alonso-Alvarez et al., 2004), because spawning is a highly demanding activity which elevates metabolic rate for an extended period of the time. At the case of single breeding event and further individual death, fish may not invest resources into generating antioxidant defense, and all resources have been allocated to reproduction as, for instance, of semelparous fish species namely salmonides (Wilson et al., 2014), while ilteroparous species are characterized by multiple breeding events throughout life.

Fish may reduce investments in their own antioxidant defense in favour of eggs and embryos antioxidant protection, which is essential for hatching success of embryos and survival of new hatching larvae (Fontagné et al., 2006; Taylor et al., 2015). As we described previously, high concentrations of low molecular weight antioxidants were detected in the gonads of Black Sea teleosts, while antioxidant enzyme activities, including CAT and PER were not high (Rudneva, 2012). Additionally, antioxidant enzyme activities were very low in fish gonads, while high concentrations of low molecular weight antioxidants play an important role in antioxidant defense of fish reproductive system against ROS and oxidative stress.

We could also found the differences in CAT and PER activity in the RBC and in the liver of tested fish male and female (Figure 7).

No significant differences of CAT and PER activities in RBC in the majority of tested fish species were observed, with the exception of red mullet *M. barbatus ponticus*, in which PER activity was significantly higher as compared with female value. At the other side, we could note, that in female enzyme activities tended to increase related to male.

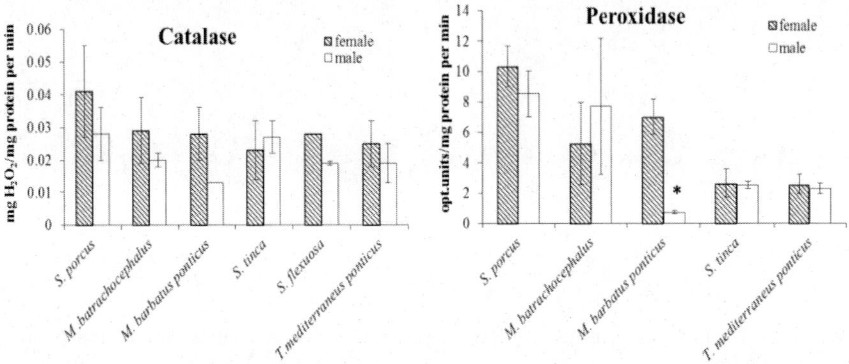

Figure 7. Gender-related differences of CAT and PER activities in RBC of Black Sea teleosts. * - significant difference.

Generally, similar trends of CAT and PER activities related to fish gender were displayed in the liver (Figure 8). However, CAT activity in *M. barbatus* and *T. mediterraneus ponticus* was significantly greater in male ($p < 0.05$).

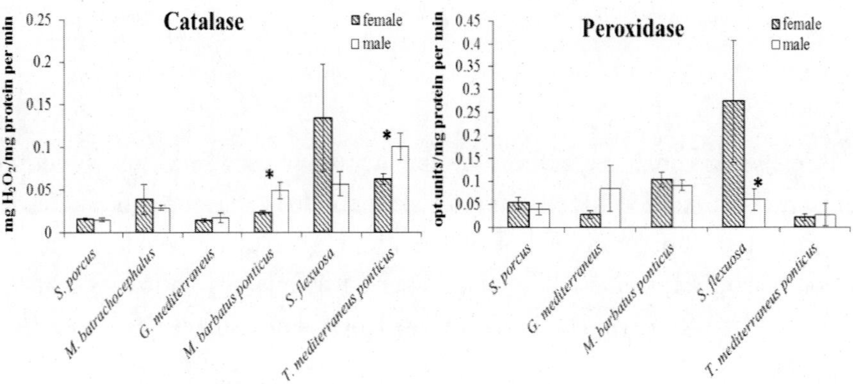

Figure 8. Gender-related differences of CAT and PER activities in the liver of Black Sea teleosts. * - significant difference.

Therefore, we could conclude, that CAT and PER level did not differ significantly in fish male and female in the majority of tested Black Sea fish species.

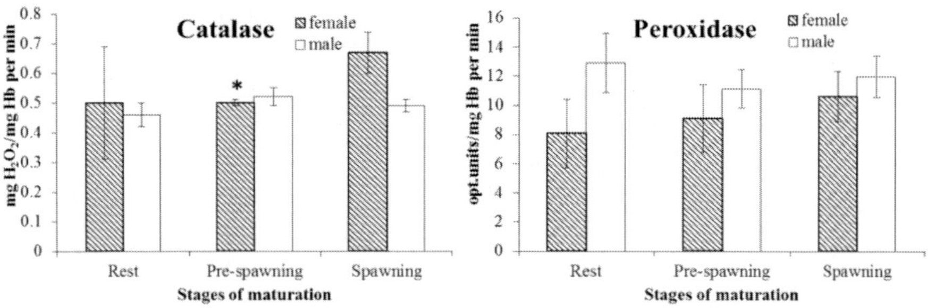

Figure 9. CAT and PER activities in RBC of *N. melanostomus* in different stages of reproductive cycle.

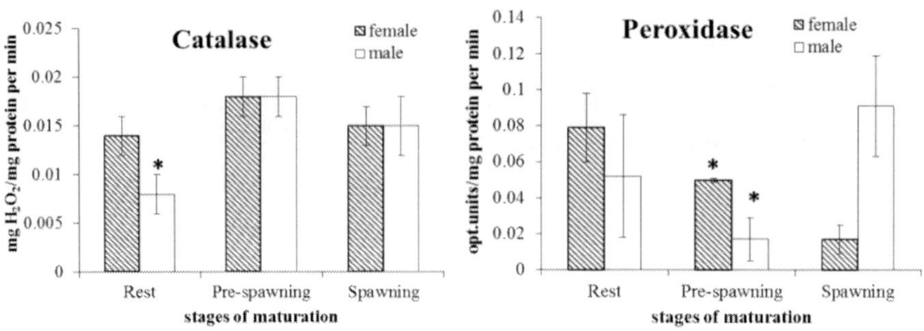

Figure 10. CAT and PER activities in the liver of *S. porcus* in different stages of reproductive cycle.

Because iteroparous species are characterized by multiple breeding events throughout their life, they are less depend to the resource-based trade-off associated with reproduction and antioxidant defense changes during maturation and reproduction of fish. There are several publications of the correlation between fish reproduction cycle and antioxidant defense status. The authors reported, that reproduction in zebra finches decreases antioxidant defense, which may suggest that oxidative stress is a cost of reproduction (Alonso-Alvarez et al., 2004). In our study we found the increase of CAT and PER activity in the RBC of the female of the round goby *N. melanostomus* in the spawning time (Figure 9). However, in male of the round goby CAT activity decreased, and we could propose that paternal care was negatively correlated with oxidative stress resistance, which agrees with the data of the other investigators (Wilson et al., 2012).

In the liver of scorpion fish *S. porcus* we showed the increase of CAT activity in pre-spawning and spawning period both in female and male, while hepatic PER level increased in male in spawning time and decreased significantly in female (Figure 10). Therefore, enzyme activities depend on the period of reproduction, fish gender, interspecies features and specificity of tissue. Additionally, the cumulative effect of oxidative stress across the reproductive lifespan of fish has been attributed with the age-associated decline of antioxidant defense. At the other side, differences in oxidative stress levels during reproduction can be correlated with the specificity of diet and food deprivation, as well as to migration of the fish species.

3.3. Diet and Food Deprivation

Antioxidant status of the organism depends on food composition, because many kinds of essential antioxidants are contained in the diet. For instance, antioxidants namely vitamin E, carotenoids are strictly obtained as a nutritional antioxidants in the majority of fish species, while the others (for instance, vitamins C, A) are synthesized only in few fish species. These components play an important role in their adaptation to the changing environment. Unlike mammals, fish may survive long time of starvation, however, without food they change their metabolism and evolved metabolic adaptations that involve the up- and down-regulation of the major biochemical and physiological pathways (Morales et al., 2004).

Antioxidant status of fish is linked with fish trophic level, feeding behavior and nutrition factors, which also may affected antioxidant enzymes including CAT and PER (Martinez-Alvarez et al., 2005; Sole et al., 2009). Fish with different diets may cope with oxidative stress in different ways. Several essential microelements, which play a role in many physiological functions, including antioxidant defenses (such as selenium, copper, iron, and etc.) fish obtain with their diet. Several of these micronutrients are the components of the active centers of the antioxidant enzymes, namely CAT (iron), PER (iron), GPxs (Se). The authors have documented, that antioxidant enzyme activities was generally higher in herbivorous fish (grass carp *Ctenopharyngodon idella* and silver carp *Hypophthalmichthys molitrix)* than in omnivorous species barbel *(Barbus barbus),* crucian carp *(Carassius carassius)* and common carp *(C. carpio)* (Radi et al., 1985). The authors suggested, that plant-based diet contains the components, which are absent in other food or the prooxidants in this type of food contains stimulated ROS production, and the change of prooxidant/antioxidant balance, attributed with the induction of enzyme activities. CAT activity was significantly higher in the liver of Senegalese sole fed high lipid diet as comparted with fish fed low lipid diets (Rueda-Jasso et al., 2004)

In our previous studies (Rudneva, 1997b; 2012) tested Black Sea teleosts species were grouped in pelagic plankton feeders (*T. mediterraneus ponticus*), omnivorous (*S. smaris*), carnivorous (*M. barbatus ponticus, N. melanostomus*) and predators (*M. merlangus euxinus, S. porcus, P. lucsus*). No clear relationships between feeding groups and CAT and PER activities in the RBC and in the liver were found. However, carnivorous fish species demonstrated approximately similar enzymatic activity in the liver. It could be explained that benthic invertebrates (mollusks, crustacean and worms) and fish, which are the preferable prey for this group might accumulate pollutants from the bottom sediments and transfer them via trophic nets to fish with the effect of concentration. Hepatic antioxidant enzyme activity in pelagic plankton feeder horse mackerel and omnivorous suprabenthic/pelagic *S. smaris* was approximately similar and lower than in carnivorous forms, which was explained the similarity of their high metabolic rate and diet consumption.

Food deprivation was also attributed with oxidative stress in fish. The researchers demonstrated the changes of prooxidant/antioxidant balance in the fish at the case of lack of food or deprivation, which was correlated with the increase of lipid peroxidation and CAT and GPx activities in common dentex (*Dentex dentex)* (Morales et al., 2004). Food consumption and oxidative stress, associated with food quality and deprivation play a role in fish migration and physical activity.

3.4. Swimming Activity

Physical activity plays an important role in antioxidant defense status in fish. Our study supports the great interspecies differences in antioxidant system in tissues between various fish species, characterizing different swimming activities. Our findings agree with the results of other researchers, who reported abot high variability of antioxidants level in the fish (Winston, 1991; Filho et al., 1993; Sole et al., 2009). Interspecies differences in blood, muscle and liver in Black Sea teleosts were more significantly than in gonads, which showed more homogenous response (Rudneva, 2012).

The interspecies variations of antioxidant defense may reflect the specific adaptations to the oxidative stress and protective mechanisms against ROS damage. Therefore, activity of antioxidant enzymes in fish blood correlated with their swimming capacity. In our previous study we found that CAT activity in fast swimming pelagic horse mackerel *T. meditrraneus ponticus* and high body pickerel *S. smaris* were significant higher than those in blood of slow swimming gobies, scorpion fish and flounder (Rudneva, 1997a, b). Other researchers reported also, that CAT content in blood of more active forms was higher as compared to more sluggish species (Filho et al., 1993; Filho, 1996). They suggested, that antioxidant defense in marine species in the liver and blood may be related to the oxygen consumption of the tissues and of the whole organism, while in freshwater fish it may be related to physical and chemical characteristics of the living conditions rather than to physical activity (Filho, 1996). At the same time, the changes in the sturgeon blood prooxidant-antioxidant status, as a consequence of adaptation to marine conditions, were not reflected in the liver and other tissues (Martinez-Alvarez et al., 2005).

Liver of the vertebrates exhibits a high metabolism and oxygen consumption, and it is the main organ of xenobiotic detoxification. Fish liver displayed the highest levels of the key antioxidant enzymes CAT (Filho et al., 1993; Filho & Boveris, 1993; Rocha-e-Silva et al., 2004). CAT level in the liver appears to indicate that the most active species of teleosts had greater enzyme activity compared with low mobile forms (Filho et al., 1993; Filho & Boveris, 1993). The higher activity of antioxidant enzymes in the liver of active fish correlated with higher oxygen consumption in fast swimming species and their high metabolic rate (Martinez-Alvarez et al., 2005; Filho, 2007). Animals with high metabolic rate exhibit the high rates of free radical production and show the induction of antioxidant defense mechanisms (Zelinski & Portner, 2000). Physical activity of fish is often associated with fitness, reproduction or predator avoidance. Therefore, the link between oxidative stress and physical activity may appear to be more complex.

Additionally, the majority of our tested teleosts migrate in different distances, while the others are settled (gobies, scorpion fish, flounder). Therefore, their antioxidant status depend on the specificity on abundance and complex of living conditions in biotope.

4. Abiotic Conditions

The researchers suggest, that oxidative stress in various aquatic organisms can correlate with several types of stressful conditions, such as chemical pollution, hypoxia/anoxia, microbial and viral infections and parasitic invasion, dietary influence, feed deprivation, acute temperature changes, algal blooms, UV-irradiation, and etc. (Ferreira Pereira, 2014). In this section we focus on the relationships between the ecological conditions in biotops and CAT and PER activities in tested teleosts.

4.1. Ecological Specificity of Abundance

Ecological conditions of the habitats play an important role in fish antioxidant status and their adaptation to habitats. Previously we described enzyme activities in various tissues in several Black Sea teleost and elasmobranch species (Rudneva, 2012). Here we focus on the main trends of CAT and PER activities in the RBC and in the liver of tested Black Sea teleosts, which grouped in four ecological classes namely, benthic, suprabenthic, suprabenthic/pelagic and pelagic. The highest CAT activity in the RBC was indicated in benthic forms *S. porcus* and *N. melanostomus,* in the species belonging to the other ecological groups the enzyme activity was comparable with the exception of *P. luscus* and *L. aurata,* which showed the low values (Figure 11). PER activity of examined teleosts did not vary uniform, and we can't find any trends in enzyme activities of species belonging to tested ecological groups. Similar observations we showed in the liver of checked teleosts, in which CAT and PER activity varied widely (Figure 12).

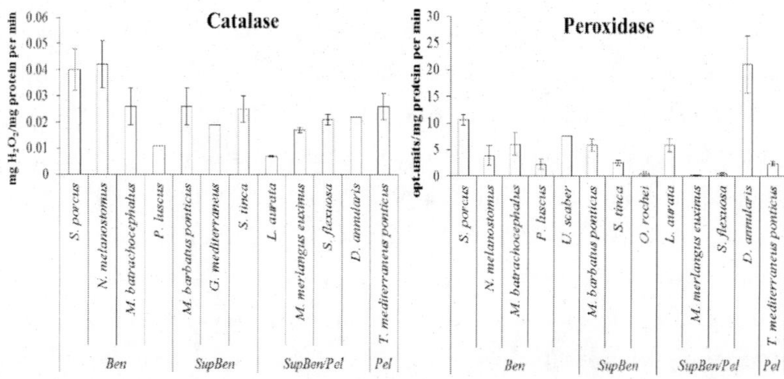

Figure 11. CAT and PER activity in RBC of Black Sea teleosts. Ecological groups: Ben – benthic, SupBen – suprabenthic, SupBen/Pel – suprabentic/pelagic, Pel – pelagic.

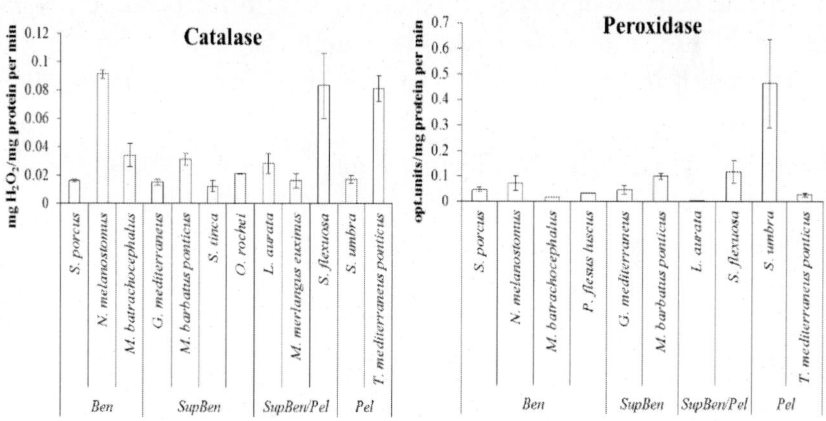

Figure 12. CAT and PER activity in the liver of Black Sea teleosts. Ecological groups: Ben – benthic, SupBen – suprabenthic, SupBen/Pel – suprabentic/pelagic, Pel–pelagic.

Therfore, the obtained results support the great interspecies differences in antioxidant activity in various tissues between examined species, which agrees with the opinion of many researchers studying fish antioxidants (Ciornea et al., 2009; Alti et al., 2016; Mukherjee et al., 2017). In our previous studies we showed the differences of low molecular weight scavengers in fish blood. For instance, glutathione content was relatively higher in pelagic active fish *T. mediterraneus ponticus* and in sluggish benthic forms, while in other species belonging to suprabenthic and suprabenthic/pelagic groups the values were intermediate (Rudneva, 2012). We found also, that SOD activity in RBC was significantly less in sluggish forms as compared to fast moving ones, which could indicate that the oxygen transport capacity of fish blood follows the same evolutionary trend of SOD level concentration in blood (Filho et al., 1993). Similar tendency was shown in CAT activity, PER level was not uniform and varied unclearly, while GR activity demonstrated the opposite trend: it was less in pelagic fish than in animals belonging to the other groups. Therefore, the results of the present study agree with our previous data.

Wide antioxidant enzyme fluctuations in the liver within the ecological group, especially in suprabenthic/pelagic class, in which enzyme activities differed in 2-7 fold, were showed in our studies. This was explained by the specificity of fish biology and life cycle related to temperature, oxygen concentration, pressure and other abiotic conditions.

Due to the highly variable environment (absorption of solar radiation, including its UV-spectra, fluctuations of salinity, depth, oxygen concentration, presence of hydrothermal vents), marine fish species impact on abiotic factors greater related to freshwater forms. These factors and their fluctuations can cause the generation of ROS.

4.2. Salinity

Due to their life history many teleost fish species were adapted to salinity fluctuations, especially in spawning period, when they migrate from freshwater rivers into marine waters (e.g., salmonids). Because salinity changes the metabolic rate of the fish associated by osmoregulatory demands and physiological stress, the ROS production changes also and fluctuations of salinity induce of oxidative stress in fish (Zeng et al., 2017). In experimental conditions the researchers indicated the increase of GPx and CAT activities in *Litopenaeus vanna*, exposed to seawater for a 48-hr period (Liu et al., 2007) and in the olive flounder *Paralichthys olivaceus* (Choi et al., 2008). Therefore, the researchers postulated, that salinity fluctuations and especially the fish transition from freshwater to marine water conditions induce oxidative stress and the changing of prooxidant/antioxidant balance. At the other side, fish respond to oxidative stress with the increase of antioxidant defense, including CAT and PER activities, and they will cope their life with changing environmental conditions.

In our studies we compared age-related trends of CAT and PER activities in the RBC of round goby *N. melanostomus* from Black Sea and Azov Sea, characterizing different salinity (Figure 13). Water salinity in the Black Sea is estimated as 18‰, while salinity in Azov Sea varies between 11.4 and 12.1‰. CAT activity in the RBC was lower in the fish from Black Sea as compared with the animals from Azov Sea. However, CAT activity decreased in the RBC of the middle-age round goby, then increased in elder ones from Black Sea, while in the animals from Azov Sea CAT activity varied lower. PER activity in the RBC of fish from Black Sea was higher related to the values of the animals from Azov Sea. It showed the opposite trend with aging as compared with CAT level, while PER activity in the RBC of round goby from Azov Sea varied insignificantly.

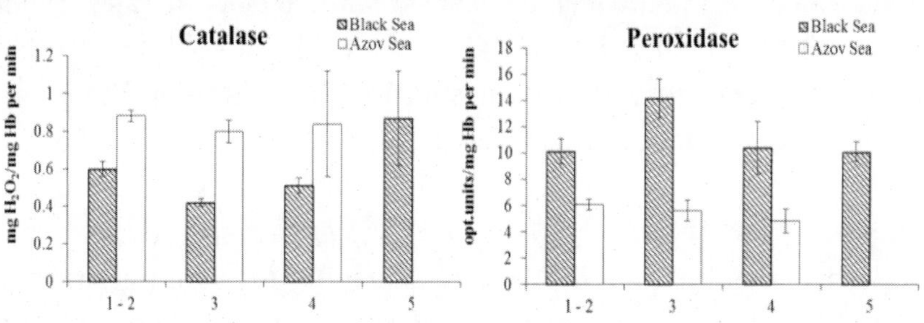

Figure 13. Age-related CAT and PER activity in RBC in *N. melanostomus* from Black Sea and Azov Sea.

Therefore, we could conclude, that CAT and PER activity of fish depends on living conditions in the abundance, especially salinity, which is the result of the evolutionary process and their adaptation to the specificity of biotope.

4.3. Pressure and Deep Sea Conditions

Deep sea biotops are characterized by reduced residual light, decreasing temperature, available nutrients, the low-oxygen environment or increased pressure. The researchers proposed, that physical and chemical conditions of this dark, poorly oxygenated environment, the meso- and bathypelagic waters of the oceans might be considered as refuges against oxidative dangers in the fish (Janssens et al., 2000). The study of antioxidants in the tissues of 16 species belonging to meso- and bathypelagic groups, which abundance was estimated between the surface and a depth of 1300 m indicated the differences of their antioxidant system status. The levels of SOD and GPx decreased in parallel with the exponential reduction in the metabolic activity in fish tissues, while CAT activity was affected neither by the metabolic activity nor by the depth of occurrence of the fishes. High levels of metabolic and antioxidant enzymes were observed in the light organs of bioluminescent species. The authors suggested that the adjustment of the activity of SOD and GPx to the decreased metabolic activity in deep-sea fish are used primarily against metabolically produced ROS, whereas the maintenance of CAT, remained unchanged while metabolic rates declined with increasing depth of occurrence (Janssens et al., 2000). The researchers postulate, that CAT and GPx activities in fish tissues were linked with the depth of their abundance (Ribata et al., 2015). Additionally, the lower metabolic rates of deep-sea animals are functionally adaptive for life in the high depth and the oxygen minimum layer and are not specifically evolved adaptations to the low food availability or low O_2 levels in these habitats (Section 4.4.). The declines of metabolic rate of the deep-sea aquatic organisms reduced metabolic requirements and this is attributed with the decrease of SOD and GPx activities. However, higher level of pollution of sediments and deep water layers is associated with the prooxidants and ROS generators (Porte et al., 2000).

In experimental conditions the researchers investigated the effects of exposure to 10.1 MPa (equivalent to 1000 m in depth) hydrostatic pressure on silver European eel (*Anguialla anguilla*, Anguillidae) oxidative stress and ROS production, no harmful oxidative damage was observed in the fish. Among our tested Black Sea teleosts we have not deep-sea forms and we can't discuss the influence of pressure on the CAT and PER enzymes of Black Sea fish.

4.4. Oxygen Availability

Marine environments are characterized by high oxygen concentrations in the euphotic zone and by a decrease to lower oxygen levels in the deep sea. Since oxygen is essential for aquatic organisms, they developed adaptations to oxygen concentrations and able to survive extreme oxygen level in deep sea regions (Janssens et al., 2000; Welker et al., 2013). In experimental conditions the researchers found an increase in fish antioxidant defense under hypoxia/anoxia, as "preparation for oxidative stress" to enhance their ability to degrade ROS production upon return to usual oxygen concentration (Hermes-Lima et al., 1998; Lushchak & Bagnyukova, 2007).

In deep-sea fish species SOD and GPx activity increased with depth, while CAT activity was not changed (Janssens et al., 2000), and the authors proposed that deep sea fish species demonstrated less oxidative stress with the decrease of their metabolic rate and low ROS production. Therefore, fish have evolved special mechanisms of regulation of prooxidant/antioxidant balance in changing oxygen conditions. Both hyperoxic and hypoxic conditions can induce oxidative stress in fish and increase antioxidant enzyme level, including CAT and GPx in tissues (Lushchak, 2011). However, oxygen availability is species-dependent, because fishes have evolved to different living conditions and they inhabit different biotops, where the oxygen level varies widely. For instance, pelagic fish species (*T. mediterraneus ponticus*) abundance is characterized high oxygen level, while benthic forms (*N. melanostomus, S. porcus*) live in the water, where oxygen concentration is lower. In addition, metabolic rate of horse mackerel is higher related to benthic forms, and we can propose, that antioxidant activity in pelagic forms is greater, than in bottom ones. In this context hepatic CAT activity in pelagic horse mackerel is significant higher, than in benthic and suprabenthic fishes, however, PER activity was very low (Figure 11). We could conclude, that characteristics of antioxidant defense in fish depended on the complex of living conditions in biotope and relationships among them and fish species, which was demonstrated at the case of temperature.

4.5. Temperature

Global climate change may cause stress to aquatic ecosystems and organisms, damage their health, growth and reproduction. Climate changes are accompanied with the temperature increasing caused the emission of greenhouse gases (CO_2, CH_4, SO_2, nitrogen oxides, and etc.). Seawater temperature of 2-3°*C* above long-term average summer temperature results stress in aquatic organisms, led oxidative damage. High temperature elicits a series of physiological responses, such as increasing of standard body metabolism. Temperature has been known to alter metabolic rate, enzyme activities, oxygen consumption and oxidative stress, attributed with the changes of prooxidant/antioxidant ratio of the organism (Dubey, 2013). It is predicted that global climate change will increase water temperature by 2°C by the end of the 21st century (IPCC Report, 2013) and biochemical alterations including oxidative stress in aquatic organisms will also elevate.

An increase in temperature is associated with an increased metabolic rate and oxygen consumption, attributed with an enhancing of ROS production (Almroth et al., 2015). The authors suggested, that although stenothermal fish can modulate their antioxidant capacity, they can only do so in a transient manner, the latter which remains insufficient in degradation of increased ROS production accompanying the enhance in metabolic rate with increased water temperature. Further accumulation of oxidative products in the cells and tissues can lead to a decrease in fitness and damage physiological functions in the organisms, which is important at the case of prediction of global warming consequences.

In experimental conditions the researchers indicated, that the activity of both SOD and CAT displayed only small changes across treatments and tissues during long-term acclimation of three notothenioid fish species to decreased seawater pH and increased temperature (Enzor & Place, 2014). The oxidative stress and antioxidant defense changes attributed with the water temperature fluctuations, were highly depended on tissues, and the researchers postulated, that basal antioxidant protection is kept high in those tissues that may have a stronger impact on health and fitness (Luschak, 2011).

We study the season changes of CAT and PER activity in the RBC and in the liver of several Black Sea teleosts. Our findings indicated, that in warm season the activity of tested enzymes increased in the majority of Black Sea teleosts, which could explain both water temperature increasing and the enhancing of fish metabolic rate (Figures 14, 15, 16).

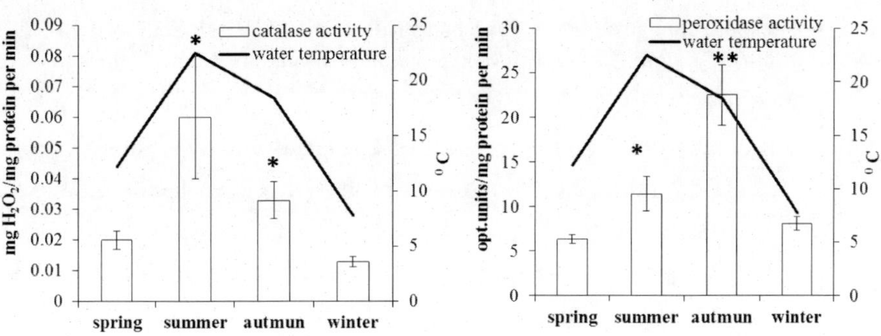

Figure 14. Seasonal changes in RBC CAT and PER activity of Black *S. porcus*.
* - significantly difference related to spring and winter values ($p < 0.05$).

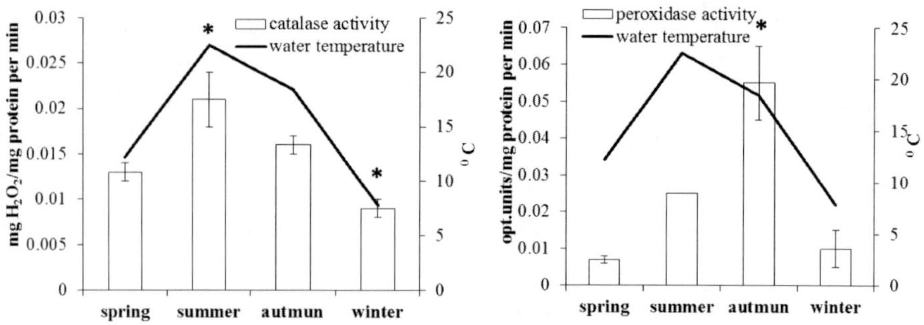

Figure 15. Seasonal changes in hepatic CAT and PER activity of Black Sea *S. porcus*. * - significantly difference related to spring and winter values ($p < 0.05$).

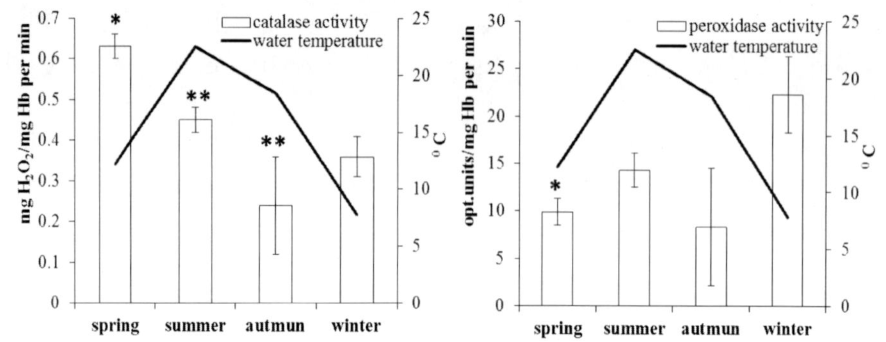

Figure 16. Seasonal changes in RBC CAT and PER activity of Black Sea *N. melanostomus*. * - significantly difference related to spring and winter values ($p < 0.05$).

In summer at high water temperature CAT activity was significantly greater in the RBC and in the liver of scorpion fish related to the values of the other seasons, while the highest PER level was indicated in autumn, when the water temperature was decreased insignificantly. In cold seasons, at the beginning of spring and winter, the enzyme activities in the RBC and in the liver of *S. porcus* was significantly lower (Figures 14, 15). However, in the RBC of round goby *N. melanostomus* high CAT activity was observed in spring and summer time, then it significantly decreased in autumn and in winter (Figure 16), while the highest PER level was indicated in winter. Our results agree with the opinion of the other investigators, who documented, that temperature-related differences of the antioxidant enzymes demonstrate tissue- and species-dependent peculiarities and evolved in the complex of fish physiological and metabolic adaptations to the living conditions (Alti et al., 2016; Mukherjee et al., 2017).

Finally, present data demonstrated that CAT and PER levels showed variations depending on the physical environment, but the changes are exclusively site as well as tissue and species specific. These findings also suggest that the effects of environmental factors on oxidative stress mechanisms may be tissue-and species-dependent and that species may have evolved different mechanisms to deal with the oxidative challenges induced by environmental stress of selection underlying adaptive divergence in these fish species. Because CAT and PER enzymes play a role in the adaptation of the fish to changing environmental conditions, they may modify their activities as a function of water temperature, pressure, salinity, oxygen concentration and the presence of various pollutants, they may use in ecotoxicological studies as biomarkers of fish response on unfavorable living conditions and anthropogenic impact.

5. Anthropogenic Impact

ROS production can be induced by external factors. Environmental pollution may be one of the strong cause of the ROS generation in the organisms. Chemicals represent in the environment (transition metals Cu and Fe, which catalyze the production of hydroxyl radicals through Fenton reaction, toxic elements Cd, Hg, Ni, biphenyls, quinones, nitroaromatics and drugs) can also induce production of superoxide by redox cycling and via biotransformation reactions (Winston, 1991; Livingstone, 2001; Lesser, 2006). The response on oxidative stress in fish is tissue-dependent and species-dependent. Fish in polluted habitats showed changes of CAT and PER activity, however, the trends were not uniform in the tissues and examined species. Finally, the results obtained provide evidence, that exposure to heavy metals in fish can unsettle the redox balance and alter antioxidant defenses. Aquatic environments are affected by various anthropogenic contaminants like pesticides, oil, polychlorinated biphenyls

(PCBs), nanoparticles, microplastic, human pharmaceuticals, which can induce oxidative stress in fish. Coastal marine ecosystems are the main recipients of waste waters and effluents, containing high concentrations of various pollutants, which modify water hydrological properties, biodiversity and community structures. The indicators of oxidative stress in fish are very sensitive to this impact and can reflect the ecological status of habitats, as we described previously (Rudneva et al., 2012; Rudneva et al., 2016a, b; 2017). The present study also demonstrates the fluctuations of CAT and PER activities as markers of antioxidant defense in fish caught in polluted and non-polluted marine areas (Figure 17). We studied CAT and PER activities in the RBC of scorpion fish collected in the Sevastopol bays with different pollution levels as following (from the most polluted to reference area): Streletskaya→Alexandrovskaya→Karantinnaya→ Kazach'ya. High CAT activity was indicated in the fish from polluted waters of Streletskaya Bay and Alexandrovskaya Bay, while in reference area Kazach'ya Bay and less polluted Karantinnaya Bay the values were lower. However, at the case of PER activity in scorpion fish RBC we found the highest level in the fish caught in Alexandrovskaya Bay, while the enzyme activity in fish RBC from the other tested bays were lower or the similar.

Furthermore, the study of CAT and PER activity in the blood of round goby from three Sevastopol bays, characterizing different levels of pollution in two periods (2003 and 2012), also demonstrated significant differences of enzyme activities (Figure 18). CAT activity was higher in the RBC of round goby in 2012 as compared with 2003 in three tested bays, similar trend was shown in PER activity with the exception of the fish from Karantinnay Bay. We could conclude, that in recent years, when anthropogenic impact on marine water was increased, the fish responded the increase of antioxidant defense, including the induction of CAT and PER level in the RBC.

At the other side, the response of antioxidant system and antioxidant enzymes can be different in the tissues and depend on fish species (Figure 19). Hepatic CAT and PER activity was higher in the majority of fish from Alexandrovskaya Bay, with the exception of horse mackerel *T. mediterraneus ponticus,* while enzyme activities in the examined fish species collected in Karantinnaya Bay were lower. Therefore, the response of CAT and PER activity in the fish on pollution impact in the tested areas was not uniform and depended both on the contamination level, specificity of the enzyme, tissues and fish species. It is very important for the understanding the main mechanisms of teleosts adaptations to changing environmental conditions, including the anthropogenic impact which is one of the main factor of the evolution at recent decades.

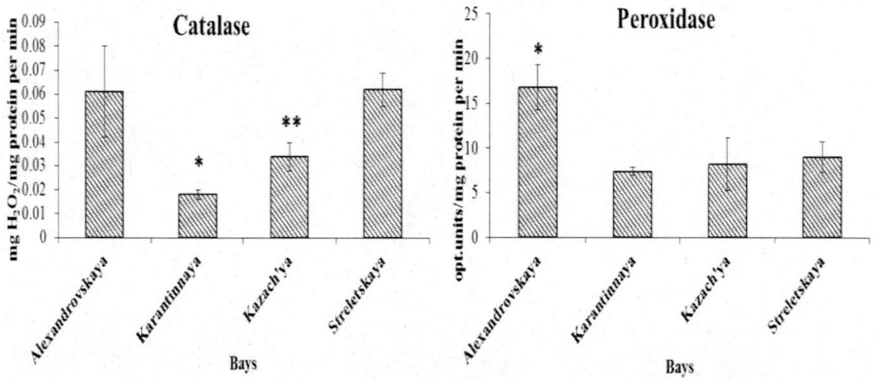

Figure 17. CAT and PER activity in RBC of *S. porcus* from different bays. * - significantly difference related to other bays ($p < 0.05$), ** - significantly difference related to Alexandrovskaya and Streletskaya bays ($p < 0.05$).

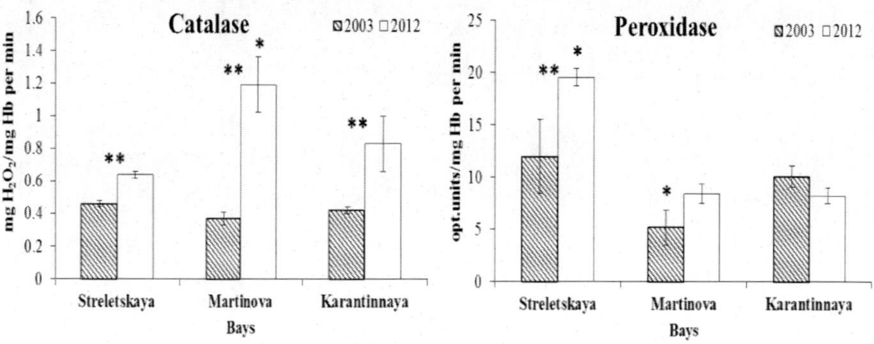

Figure 18. CAT and PER activity in RBC of *N.. melanostomus* from different bays at the period of 2003 and 2012. * - significantly difference related to other bays ($p < 0.05$), ** - significantly difference related to periods ($p < 0.05$).

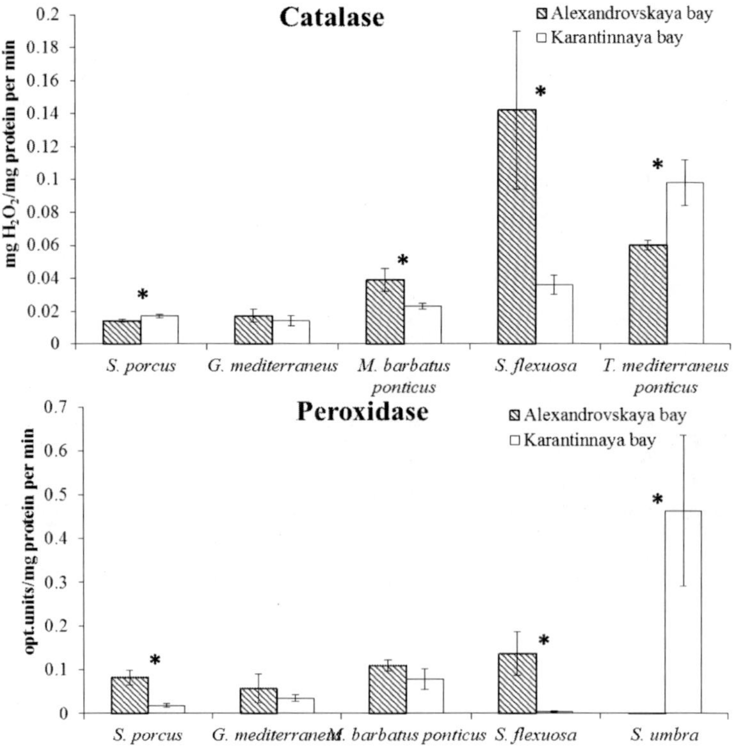

Figure 19. CAT and PER activity in the liver of Black Sea teleosts from different bays. * - significantly difference between tested bays ($p < 0.05$).

These findings suggest that in general, pollution of aquatic environments reflected physiologically in fish by perturbing normal free radical processes, leading to increase in oxidative damage and disturbance of antioxidant defenses. It would therefore be advisable to consider oxidative ecology in fisheries science and monitoring, given the extent of pollution in many waterbodies and the fish species that live within them (Birnie-Gauvin et al., 2017).

In recent decades, the new pollutants, namely microplastic, nanoparticules and human pharmaceuticals, are highly distributed in the aquatic environments and they are accumulated in the animals' tissues. The information of their biological effects is very limited and there is an increasing need to assess wild populations' subtle alterations that may indicate the presence of environmental stressors (Amerand et al., 2010; Oliveira et al., 2017a, b). Antioxidants and key antioxidant enzymes CAT and PER may be used for these purposes as a good biomarkers.

CONCLUSION

On the basis of the data described above, we can conclude, that key antioxidant enzymes catalase and peroxidase in tested Black Sea teleosts may provide information about interspecies differences and the mechanisms of their adaptation to environmental conditions, including the complex of biotic, abiotic and anthropogenic factors. Moreover, the global climate changes in recent decades can directly and indirectly influence on fish antioxidant status and together with the local factors may be suitable for the understanding of the mechanisms of fish evolution and adaptation to variety of ecological conditions in marine ecosystems. At recent decades, anthropogenic impact on marine environment has become one of the important factor of ecosystem transformation and could use in the evolutionary process of fish. Therefore, fish are good monitors for the understanding of this process and their biomarkers may successfully use in monitoring programs.

REFERENCES

Aceto, A., Amicarelli, F., Sacchetta, P., Dragani, B., Bucciarelli, T., Masciocco, L., Miranda, M. & Di Ilio, C. (1994). Developmental aspects of detoxifying enzymes in fish (*Salmo iridaeus*). *Free. Rad. Res*., 21 (5), 285-294.

Ahmed, R. G. (2005). Is there a balance between oxidative stress and antioxidant defense system during development? *Medical J. of Islamic World Academy of Sciences,* 1 (2), 55-63.

Almroth, B. C., Asker, N., Wassmur, B., Rosengren, M., Jutfelt, F., Gräns, F. & Sturve, J. (2015). Warmer water temperature results in oxidative damage in an Antarctic fish, the bald notothen. *Journal of Experimental Marine Biology and Ecology*, 468, 130-137.

Alonso-Alvarez, C., Bertrand, S., Devevey, G., Prost, J., Faivre, B. & Sorci, G. (2004). Increased susceptibility to oxidative stress as a proximate cost of reproduction. *Ecology Letters*, 7, 363-368.

Alti, G., Canli, E. G., Eroglu, A. & Canli, M. (2016). Characterization of antioxidant system parameters in four freshwater fish species. *Ecolotox. Environ. Safety.*, 126, 30-37.

Amérand, A., Vettier, A., Moisan, C., Belhomme, M., & Sébert, P. (2010). Sex-related differences in aerobic capacities and reactive oxygen species metabolism in the silver eel. *Fish physiology and Biochemistry*, 36, 741-747.

Birnie-Gauvin, K., Costantini, D., Cooke, S. J. & Willmore, W. G. (2017). A comparative and evolutionary approach to oxidative stress in fish: A review. *Fish and Fisheries*, 18 (5), 928-942.

Boltachev, A. R. & Karpova, E. P. (2017). *Marine fishes of the Crimean peninsula*. (2nd edition). Simferopol: "Biznes-Inform." (in Russian).

Cao, L., Huang, W., Liu, J., Yin, X. & Dou, Sh. (2010). Accumulation and oxidative stress biomarkers in Japanese *flounder* larvae and juvenile sunder cadmium exposure. *Comp. Biochem. Physiol.,* 151C, 386-392.

Choi, C. Y., An, K. W. & An, M. I. (2008). Molecular characterization and mRNA expression of glutathione peroxidase and glutathione-S-transferase during osmotic stress in olive Flounder (*Paralichthys olivaceus*). *Comp. Biochem. Physiol.*, 149A, 330-337.

Ciornea, E., Vasile, G., Cojocaru, D. & Cojocaru, S. I. (2009). A comparative study on the activity of hepatic and muscular catalase in freshwater fish species. *Analele Stiintifice ale Universitatii "Alexandru Ion Cuza," Sectiunea Genetica si Biologie molculara*, 10, 13-18.

De La Fuente, M., Hermanz, A. & Vallejio, M. C. (2005). The Immune System in the Oxidative Stress Conditions of Aging and Hypertension: Favorable Effects of Antioxidants and Physical Exercise. *Antioxidnts & Redox Signaling*, 7 (9-10), 1356-1366.

Dubey, A. K. (2013). Temperature stress induces neuronal damage via a mechanism leading to free radical production and altering endogenous antioxidant defences in freshwater Indian catfish *Heteropneustes fossilis. The Journal of Free Radicals and Antioxidants*, 139, 204-209.

Enzor, L. A. & Place, S. P. (2014). Is warmer better? Decreased oxidative damage in notothenioid fish after long- term acclimation to multiple stressors. *J. Exp. Biol.*, 217, 3301-3310.

Falkowski, P. G., Katz, M. E., Knoll, A. H., Quigg, A., Raven, J. A., Schofield, O. & Taylor, J. R. (2004). The evolution of modern eukaryotic phytoplankton. *Science*, 305, 354-360.

Ferreira Pereira, L. F. (2014). *"Growth performance, Antioxidant and innate immune responses in European seabass fed probiotic supplemented diet at three rearing temperatures"*. Thesis, University Porto, 72 pp.

Filho, W. D. (1996). Fish antioxidant defenses: a comparative approach. *Braz. J. Med. Biol. Res.,* 29 (12), 1736-1742.

Filho, W. D. (2007). Reactive oxygen species, antioxidants and fish mitochondria. *Fishery in Bioscience*, 12, 1229-1237.

Filho, D. W. & Boveris, A. (1993). Antioxidant defenses in marine fish. II. Elasmobranchs. *Comp. Biochem. Physiol.*, 106C, 415-418.

Filho, D. W., Giulivit, C. & Boveris A. (1993). Antioxidant defences in marine fish. – I. Teleosts. *Comp. Biochem. Physiol.*, 106C (2), 409-413.

Fisher, S. A. & Burggren, W. W. (2007). Role of hypoxia in the evolution and development of the cardiovascular system. *Antioxidants & Redox signaling*, 9 (9), 1339-1352.

Fontagné, S., Bazin, D., Brèque, J., Vachot, C., Bernarde, C., Rouault, T. & Bergot, P. (2006). Effects of dietary oxidized lipid and vitamin A on the early development and antioxidant status of Siberian sturgeon (*Acipenser baeri*) larvae. *Aquaculture*, 257, 400-411.

Harman, D. (2003). The Free Radical Theory of Aging. *Antioxidnts & Redox Signaling,* 5 (5), 557-561.

Hermes-Lima, M., Storey, J. M. & Storey, K. B. (1998) Antioxidant defenses and metabolic depression. The hypothesis of preparation for oxidative stress in land snails. *Comp. Biochem. Physiol.*, 120B, 437-448.

IPCC, 2013: Climate Change 2013: The Physical Science Basis. In: T. F. Stocker, D. Qin, G. K. Plattner, M. Tignor, S. K. Allen, J. Boschung, A. Nauels, Y. Xia, V. Bex & P. M. Midgley (Eds.), *Contribution of*

Working Group I to the Fifth Assessment Report of the Intergovernmental Panel on Climate Change. United Kingdom and New York, Cambridge: Cambridge University Press; 2013, 1535 pp.

Janssens, B. J., Childress, J. J., Baguet, F. & Rees, J.-F. (2000). Reduced enzymatic antioxidative defense in deep-sea fish. *J. Exp. Biol.*, 203 (24), 3717-3725.

Kalaimani, N., Chakravarthy, N., Shanmugham, R., Thirunavukkarasu, A. R., Alavandi, S. V. & Santiago, T. S. (2008). Anti-oxidant status in embryonic, post-hatch and larval stages of Asian sea bass (*Lates calcarifer*). *Fish Physiol. Biochem*., 34, 151-158.

Lau, P. S. & Wong, H. L. (2003). Effect of size, tissue and location on six biochemical markers in the green–lipped mussel, *Perna viridis*. *Mar. Pollut. Bull.,* 46, 1563-1572.

Lesser, M. P. (2006). Oxidative stress in Marine environment: Biochemical and Physiological Ecology. *Annu. Rev. Physiol*, 68, 253-278.

Liu, Y., Wang, W., Wang, A., Wang, J. & Sun, R. (2007). Effects of dietary vitamin E supplementation on antioxidant enzyme activities in *Litopenaeus vannamei* (Boone, 1931) exposed to acute salinity changes. *Aquaculture*, 265, 351-358.

Liu, W., Zhao, Ch., Wang, P., Wang, Sh., Lin, H. & Qiu, L. (2018). The response of glutathione peroxidase 1 and glutathione peroxidase 7 under different oxidative stresses in black tiger shrimp, *Penaeus mono*. *Comp. Biochem. Physiol*., 217, 1-13.

Livingstone, D. R. (2001). Contaminant-stimulated reactive oxygen species production and oxidative damage in aquatic organisms. *Mar. Pollut. Bull.*, 42, 656-665.

Lushchak, V. I. (2011). Environmentally induced oxidative stress in aquatic animals. *Aquat. Toxicol.*, 101, 13-30.

Lushchak, V. I. & Bagnyukova, T. V. (2007). Hypoxia induces oxidative stress in tissues of a goby, the rotan *Perccottus glenii*. *Comp. Biochem. Physiol.*, 148B, 390-397.

Martinez-Alvarez, R. M., Morales, A. E. & Sanz, A. (2005). Antioxidant defenses in fish: biotic and abiotic factors. *Reviews in Fish Biology and fisheries*, 15, 75-88.

Morales, A. E., Perez-Jimenez, A., Hidalgo, M. C., Abellian, E. & Cardenete, G. (2004). Oxidative stress and antioxidant defenses after prolonged starvation in *Dentex dentex* liver. *Comp. Biochem. Physiol.*, 139C, 153-161.

Mourente, G., Tocher, D. R., Diaz, E., Grau, A. & Pastor, E. (1999). Relationships between antioxidants, antioxidant enzyme activities and lipid peroxidation products during early development in *Dentex dentex* eggs and larvae. *Aquaculture,* 179 (1-4), 309-324.

Mukherjee, J., Moniruzzaman, M., Chakraborty, S. B., Lek, S. & Ray, S. (2017). Towards a physiological response of fishes under variable environmental conditions: an approach through neutral network. *Ecological Indicators*, 78, 381-394.

Nayak, S. B., Jena, B. S. & Patnaik, B. K. (1999). Effects of age and manganese (II) chloride on peroxidase activity of brain and liver of the teleost, *Channa punctatus*. *Exp. Gerontol.*, 34, 367-374.

Oliveira, M., Franco L., Balasch, J. C., Fierro-Castro, C., Tvarijonavicinte, A., Soares, A. M. V. M., Tort, L. & Teles, M. (2017a). Tools to assess effects of human pharmaceuticals in fish: a case study with gemfibrozil. *Ecological Indicators*. 29 Dec. 2017. Available from: URL: https://www.sciencedirect.com/science/article/pii/S1470160X17 308415

Oliveira, M., Tvarijonaviciute, A., Trindade, T. Soares, A. M. V. M., Tort, L. & Teles, M. (2017b). Can non-invasive methods be useful to assess effects of nanoparticles in fish? *Ecological Indicators*. 24 June 2017. Available from: URL: https://www.sciencedirect.com/science/article/ pii/S1470160X17303667?via%3Dihub.

Otto, D. M. E. & Moon, T. W. (1996). Endogenous antioxidant systems of two teleost fish, the rainbow trtout and the black bullhead, and the effect of age. *Fish Physiol. Biochem.*, 15, 349-358.

Øvergård, A. C., Eichner, C., Nilsen, F. & Dalvin, S. (2017). Molecular characterization and functional analysis of a salmon louse (*Lepeophtheirus salmonis*, Krøyer 1838) heme peroxidase with a potential role in extracellular matrixes. *Comp. Biochem. Physiol.*, 206, 1-10.

Peters, L. D. & Livingstone, D. R. (1996). Antioxidant enzyme activities in embryologic and early larval stages of turbot. *J. Fish Biol.,* 49, 986-997.

Porte, C., Escartín, E., García, L. M., Solé, M. & Albaigés, J. (2000). Xenobiotic metabolising enzymes and antioxidant defences in deep-sea fish: relationship with contaminant body burden. *Mar. Ecol. Prog. Ser.,* 192, 259-266.

Radi, A. A. R., Hay, D. Q., Matokovics, B. & Gabrielak, T. (1985). Comparative antioxidant enzyme study in freshwater fish with different types of feeding behaviour. *Comp. Biochem. Physiol.*, 81C, 395-399.

Ribalta, C., Sanchez-Hernandez, J. C. & Sole, M. (2015). Hepatic biotransformation and antioxidant enzyme acivities in Mediterranean fish from different habitat depths. *Science Total Environ.,* 532, 176-183.

Rocha-e-Silva, T. A. A., Rossa, M. M., Rantin, F. T., Matsumura-Tundisi, T., Tundisi, J. G. & Degterev, I. A. (2004). Comparison of liver mixed-function oxygenase and antioxidant enzymes in vertebrates. *Comp. Biochem. Physiol.*, 137C, 155-165.

Rudneva, I. I. (1997a). Blood antioxidant system of Black Sea elasmobranch and teleosts. *Comp. Biochem. Physiol.,* 118C, 255-260.

Rudneva, I. I. (1997b). Ecological and phylogenetic peculiarities of antioxidant enzyme activities and antioxidant content in Black Sea teleosts and elasmobranch. *J. Evol. Biochem. Physiol.*, 33, 29-37. (*in Russian*).

Rudneva, I. I. (1999). Antioxidant system of Black Sea animals in early development. *Comp. Biochem. Physiol.*, 122C, 265-271.

Rudneva, I. I. (2012). Antixidant defence in marine fish and its relationship to their ecological status. In: S. P. Dempsy (Ed.), *Fish Ecology*. (pp. 31-59). New York: Nova Science Publishers.

Rudneva, I. (2013). *Biomarkers for stress in fish embryos and larvae*. Boca Raton, USA: CRC Press.

Rudneva, I. I., Kuzminova, N. S., Skuratovskaya, E. N. & Kovyrshina T. B. (2010a). Age composition and antioxidant enzyme activities in blood of Black Sea teleosts. *Comp. Biochem. Physiol.*, 151C, 229-239.

Rudneva, I. I., Kuzminova, N. S., Skuratovskaya, E. N. & Kovyrshina, T. B. (2010b). Comparative study of glutathione-S-transferase activity in tissues of some Black Sea teleosts. *Int. J. Sci. Nat.*, 1 (1), 1-6.

Rudneva, I. I., Skuratovskaya, E. N., Dorohova, I. I. & Kovyrshina, T. B. (2012). Use of fish blood biomarkers for evaluating of marine environment health. *World J. Sci. Technol.*, 2 (7), 19-25.

Rudneva, I. I., Skuratovskaya, E. N., Chesnokova, I. I., Shaida, V. G. & Kovyrshina T. B. (2016a). Biomarker response of Black Sea scorpion fish *Scorpaena porcus* to anthropogenic impact. In: A. Kovács & P. Nagy (Eds.), *Advances in Marine Biology*. (pp. 119-147). New York: Nova Science Publishers.

Rudneva, I. I., Skuratovskaya, E. N., Chesnokova, I. I., Kovyrshina, T. B., Omelchenko, S. O. & Zalevskaya, I. N. (2016b). Fish Biodiversity: The Role of Defense Systems to the Adaptation to Ecological Conditions in the Black Sea Ecosystem. In: A. Mills (Ed.), *Baltic and Black Sea: Ecological Perspectives, Biodiversity and Management*. (pp. 111-146). New York: Nova Science Publishers.

Rudneva, I. I., Rudyk, M. P., Shepelevich, V. V., Skivka, L. M., Roslova, N. N., Skuratovskaya, E. N., Chesnokova, I. I., Shaida, V. G. & Kovyrshina, T. B. (2017). Response of marine fish liver on environmental pollution. In: G. E. Zaikov, L. I. Weisfeld, E. M. Lisitsyn & S. A. Bekuzarova (Eds.), *Heavy Metals and Other Pollutants in the Environment. Biological Aspects*. (pp. 383-401). Waretown; Boca Raton: Apple Academic Press; CRC Press, USA.

Rueda-Jasso, O. R., Dias L. E. C., De Coen, J., Gomes, W., Rees, E. J. F. & Sorgeloos, P. (2004). Effect of dietary non- protein energy levels on condition and oxidative status of Senegalese sole (*Solea senegalensis*) juveniles. *Aquaculture*, 231, 417-433.

Sole, M., Portrykus, J., Fernandez-Diaz, C. & Blasco, J. (2004). Variations on stress defenses and metallothionein levels in the Senegal sole, *Solea senegalensis*, during early larval stages. *Fish Physiol. Biochem.,* 30, 57-66.

Sole, M., Rodriguez, S., Papiol, V., Maynou, F. & Cartes J. E. (2009). Xenobiotic metabolism markers in marine fish with different trophic strategies and their relationship to ecological variables. *Comp. Biochem. Physiol.*, 149C (1), 83-89.

Semisei, I., Rao, G. & Richardson, A. (1989). Changes in the expression of superoxide dismutase and catalase of age and dietary restrictions. *Biochem. Biophys. Res. Commun.*, 165, 620-625.

Takahashi, A., Totsuni-Nakano, H., Nakano, M., Mashiko, S., Suzuki, N., Ohma, C. & Inaba, H. (1989). Generation of O_2 and tyrosine cation-mediated chemiluminescence during the fertilization of sea urchin eggs. *FEBS Lett.*, 246 (1-2), 117-119.

Taylor, J. J., Wilson, S. M., Sopinka, N. M., Hinch, S. G., Patterson, D. A., Cooke, S. J. & Willmore, W. G. (2015). Are there intergenerational and population-specific effects of oxidative stress in sockeye salmon (*Oncorhynchus nerka*). *Comp. Biochem. Physiol.*, 184A, 97-104.

Welker, A. F., Moreira, D., Campos, E. G. & Hermes-Lima, M. (2013). Role of redox metabolism for adaptation of aquatic animals to drastic changes in oxygen availability. *Comp. Biochem. Physiol.*, 165A, 384-404.

Wilson, S. M., Gravel, M., Mackie, T. A., Willmore, W. G. & Cooke, S. J. (2012). Oxidative stress associated with paternal care in small mouth bass (*Micropterus dolomieu*). *Comp. Biochem. Physiol.*, 162A, 212-218.

Wilson, S. M., Taylor, J. J., Mackie, T. A., Patterson, D. A., Cooke, S. J. & Willmore, W. G. (2014). Oxidative stress in Pacific salmon (*Oncorhynchus* spp.) during spawning migration. *Physiol. Biochem. Zool.*, 87, 346-352.

Winston, G. W. (1991). Oxidants and antioxidants in aquatic organisms. *Comp. Biochem. Physiol.*, 100C (1-2), 173-176.

Yang, D., Guo, X., Xie, T. & Luo, X. (2018). Reactive oxygen species may play an essential role in driving biological evolution: The Cambrian Explosion as an example. *J. Environ. Sci.*, 63, 218-226.

Yargicoglu, P., Agar, A., Gumuslu, S. & Oguz, Y. (1999). Age-related alterations in antioxidant enzymes, lipid peroxide levels and somatosensory-evoked potentials: effect of sulfide dioxide. *Arch. Environ. Contam. Toxicol.*, 37, 554-560.

Zámocký M., Gasselhuber B., Furtmüller P. G., Obinger Ch. (2014). Turning points in the evolution of peroxidase–catalase superfamily: molecular phylogeny of hybrid heme peroxidases. *Cell. Mol. Life Sci.*, 71, 4681–4696.

Zelinski, S. & Portner, H. O. (2000). Oxidative stress and antioxidative defense in cephalopods: a function of metabolic rate or age? *Comp. Biochem. Physiol.,* 125B, 147-160.

Zeng, L., Ai, Ch. X., Wang, Y. H., Zhang, J. Sh. & Wu, Ch. W. (2017). Abrupt salinity stress induces oxidative stress via the Nrf2-Kep1signaling pathway in large yellow croaker *Pseudosciaena crocea. Fish Physiol. Biochem.,* 43 (4), 955-964.

In: Teleosts
Editor: Mikael Herleif

ISBN: 978-1-53613-660-9

Chapter 2

The Ratio of Constitutive and Reparative Neurogenesis in Pallium of Juvenile Masu Salmon (*Oncorhynchus masou*)

***E. V. Pushchina*[1,2], *E. I. Zharikova*[1] and *A. A. Varaksin*[1]**

[1]National Scientific Center of Marine Biology, Far Eastern Branch of the Russian Academy of Sciences, Vladivostok, Russia

[2]A. A. Bogomoletz Institute of Physiology, National Academy of Sciences of Ukraine, Kiev, Ukraine

Abstract

The chapter considers the ratio of constitutive and reparative neurogenesis in pallium of juvenile masu salmon *Oncorhynchus masou*. Since salmonid juveniles are characterized by a high level of constitutive neurogenesis, we investigated the proliferative activity and neurodifferentiation in the dorsal pallial area of juvenile masu salmon. The study of proliferative activity in pallium of juvenile masu salmon *O. masou* has allowed us to identify the superficially located periventricular proliferative zone, which corresponds to the pallial periventricular zone

(PVZ) of other fish species, including the dorsal, lateral, and medial compartments. The PCNA+ cells are also identified in parenchyma of masu salmon intact brain, and their maximum concentration is observed in the medial zone. In intact brain, solitary proliferating cells were identified in parenchymal regions with their maximum concentration in the medial zone. After a mechanical injury, the zones of reparative neurogenesis — neuroepitelial neurogenic niches and zones of reactive neurogenesis surrounded by radial glial fibers—appear in masu salmon pallium. An increase in the proliferative activity is characteristic of all compartments of the PVZ (especially of the medial zone), including its deep parenchymal layers. The PVZ of juvenile masu salmon contains clusters of undifferentiated HuCD+ neurons. A change in the HuCD+ cell topography is observed in the mechanically injured masu salmon pallium, in particular, reactive neurogenic niches in the lateral zone and an increase in the cell distribution density and cell migration patterns in the medial zone. A high level of constitutive neurogenesis is characteristic of juvenile masu salmon brain. Undifferentiated HuCD+ neurons form clusters in the PVZ of intact juvenile masu salmon brain. Neurons at different stages of maturation, differing in the level of HuCD activity, were detected in the parenchymal layers of the dorsal region. A mechanical injury of masu salmon pallium is accompanied by a change in the topography of HuCD+ cells. Formation of reactive neurogenic niches is characteristic of the lateral zone, while the medial zone displays an increase in the density of immunopositive and negative cells, as well as a distinct pattern of cell migration. HuCD and PCNA immunolabeling patterns in intact pallium suggest intensive constitutive neurogenesis in the pallial PVZ of juvenile masu salmon. After a mechanical injury, the neurogenic activity spreads to deeper layers of parenchyma. The main sources of new neurons during the reparative neurogenesis are reactive neurogenic niches.

INTRODUCTION

Neuronal regeneration is the restoration of structural integrity of injured cells and fibers; it is limited by the CNS ability to produce new cells at an adult age (Kempermann, 2013). Bony fishes is a group of vertebrates that retain through adulthood the ability to produce new cells in the specialized brain regions, proliferative zones (Lema et al., 2005; Lindsey et al., 2007), which, unlike mammals, are present in many brain regions. The periventricular brain region displays the highest proliferative activity during embryogenesis, while the active neurogenesis in adults resides in the secondary matrix regions of the brain located near the external walls of different brain regions: optic tectum, cerebellum, and myelencephalon at different levels from mesencephalic isthmus to rhombencephalon (Kaslin et al., 2008).

Unlike other vertebrates, a specific everted type of hemisphere formation and development is characteristic of telencephalon in bony fishes: the intracerebral ventricles are absent, while the matrix zones are located near the external walls of the dorsal and ventral zones and are able to produce new cells both under normal conditions and after injuries (Dirian et al., 2014). The proliferative activity in pallium of adult *Danio rerio* concentrates in the dorsal proliferative zone, which corresponds to the pallial periventricular zone (PVZ), and the ventral zone, corresponding to the subpallial PVZ (Adolf et al., 2006). The preliminary studies have shown the importance of phylogenetic studies of the pallial and subpallial proliferative zones in the forebrain of the zebrafish and other fish species. Lindsey et al. (2014) demonstrated the presence of a glial-like phenotype of neural stem cells (NSCs) involved in the adult neurogenesis and the expression of various transcription factors. According to other data, PVZs contain slowly and rapidly proliferating cells generated by adult NSCs/precursor cells with radial glial-like phenotype (Adolf et al., 2006). Neuroepitelial progenitors were also recently discovered in the adult zebrafish pallium (Dirian et al., 2014). A major part of the adult pallial ventricular zone is occupied by radial glia that acts as constitutive NSCs.

The neurogenesis in adult vertebrate brain can be modulated by changing the level of cell proliferation, their survival, and/or differentiation. This is associated with different functional requirements to new neurons, such as provision of specialized functions related to sexual behavior during the mating season in songbirds (Kaslin et al., 2008; Kempermann, 2013) and emergence of new neurons in the reticular formation involved in the innervation of salmonid spinocaudal muscles, increasing in volume during individual development (Pushchina et al., 2012).

According to the lesion paradigm formulated by Zupanc et al. (1998), a high regenerative potential of the fish CNS is determined by a set of processes, including the CNS response to injury. It is currently known that the insight into regeneration-competent organisms gives a certain advantage in studying the biology of NSCs, their interaction with the cell microenvironment, and specific features of their metabolism (Berninger et al., 2006; Stocum, 2006). The neurogenesis in an adult mammalian and human brain takes place only in the subventricular zone of the lateral ventricular wall and in the subgranullar zone of the hippocampal dentate formation (Chojnacki et al., 2009; Kaneko and Sawamoto, 2009). Neurogenesis can be stimulated or inhibited by various factors, including physical activity (Kempermann, 2011), changes in environmental conditions (Lema et al., 2005), and stress (Hutton et al., 2015). These factors have multidirectional effects on the intensity of neurogenesis and determine the final success of this process.

Variations in the physiological status of the animal, beyond regenerative conditions, are important players controlling the activation state of germinal zones in the mammalian forebrain. These aspects remain understudied in fish brain. The lack of standardized animal maintenance conditions explains the large degree of variability between animals that complicates reading of subtle effects. Several neurotransmitters and hormones were recently shown to impact on radial glia activation (Mouriec et al., 2009; Perez et al., 2013). In the paraventricular organ for example, serotonine (5HT) is necessary for proliferation of radial progenitors (Perez et al., 2013), and the projection of 5HT processes toward pallial germinal zones suggests that it may be the case in this location as well (Lillesaar et al., 2009). Aromatase B (AroB), which converts androgens to estrogen, is strongly expressed in zebrafish radial glia (Menuet et al., 2005). Blockage of the aromatase activity in adult zebrafish led to a trend increase in the number of proliferating progenitors in several brain areas, particularly in pallium, suggesting a lower proliferation of oestrogens (Diotel et al., 2013). However, it remains to be assessed whether they primarily target radial glia or nonglial proliferating progenitors. Along with a quiescence-promoting effect on radial glia, radial glia, maintaining AroB expression after injury, are not activated and presumably do not participate in the repair process (Diotel et al., 2013).

A study of the constitutive neurogenesis under natural conditions gives a large volume of information for comparative analysis (Than-Trong and

Bally-Cuif, 2015). However, the balance between constitutive and injury-induced neurogenesis in brain of regeneration-competent organisms is rather vague. Recent works considered the effect of behavioral challenges or sensory stimuli on zebrafish adult neurogenesis (Lindsey and Tropepe, 2014). As the most interesting fact, it was found that sensory stimulation (chemosensory or visual stimulations) selectively affected the neurogenic niches in the brain areas involved in processing these stimulations (Lindsey et al., 2014). Changes in the social context, such as social isolation or social novelty, also decrease the number of proliferating progenitors in sensory niches and increase the number of newborn neurons (Lindsey and Tropepe, 2014). The number and complexity of neurogenic niches in teleosts, and the complex repertoire of behaviors that a fish can exhibit, suggest that much is to be learned from this model in terms of the environmental modulation of adult stem cell pools and neurogenesis. The goal of our chapter was to observe the proliferation and neurogenesis in pallium of juvenile masu salmon *Oncorhynchus masou* before and after a mechanical injury.

PCNA Localization in the Masu Salmon Pallial PVZ in Norm and after a Mechanical Injury

PCNA was immunolocalized in the dorsal, lateral, and medial zones of pallium in juvenile masu salmon (Pushchina et al., 2017). The PCNA+ cell distribution patterns in these zones were somewhat similar: solitary proliferating cells or their small clusters were observed in the surface and subventricular layers (Figure 1A, C, E). The main specific feature in the distribution of immunopositive cells in the intact animals was the pronounced surface periventricular layer containing PCNA+ conglomerates of cells that spread over all zones of the dorsal region (Figure 1A, C, E). Another specific feature consisted in the presence of solitary PCNA+ cells in deep parenchymal layers beyond the surface proliferative zone (Figure 1A, C, E). Similar patterns of PCNA+ cell distribution have been observed in other fish species (Lindsey et al., 2007; Pushchina et al., 2007; Zupanc and Sîrbulescu, 2013). Solitary PCNA+ cells were observed in the dorsal and lateral zones (Figure 1A, C) of masu salmon pallium versus the medial zone with more abundant immunopositive cells (Figure 1E).

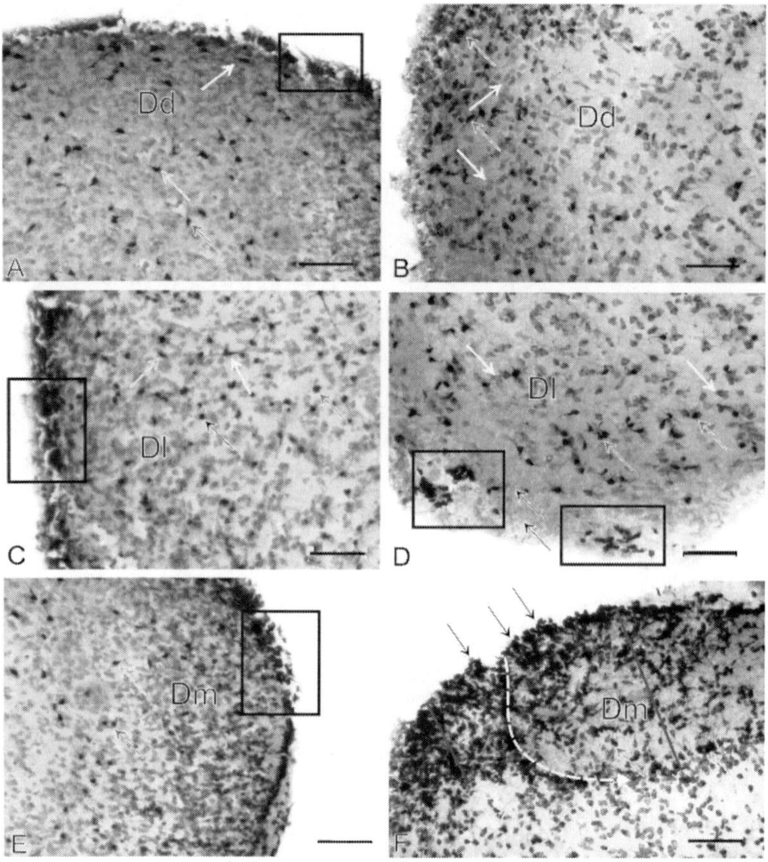

Figure 1. Distribution of the proliferating cell nuclear antigen (PCNA) in the pallial proliferative zone and deep layers of the dorsal telencephalon in juvenile masu salmon *Oncorhynchus masou* (A, C, E) in the norm and (B, D, and F) after mechanical injury: (A) dorsal (Dd), (C) lateral (Dl), and (E) medial (Dm) zones of the masu salmon dorsal telencephalon in the control (dashed line shows the boundaries of the proliferative zone; black arrows indicate solitary PCNA+ cells in the PVZ; red rectangle encloses PCNA+ cell clusters in the dorsal and medial zones; and red arrows indicate PCNA+ cells in parenchyma) and (B) dorsal, (D) lateral, and (F) medial zones after a mechanical injury (ovals enclose neurogenic niches; rectangle in panel (F), the zone of secondary neurogenesis; red arrows indicate PCNA+ cells in parenchyma and in panel (F), in the periventricular proliferative zone; yellow arrows indicate radial glial fibers; white arrows indicate clusters of migrating cells, and black arrows indicate PCNA– cells in parenchyma). Scale bar is 100 μm.

Earlier studies demonstrated that the proliferative activity in telencephalon of adult *D. rerio* usually concentrated in the dorsal proliferative zone, corresponding to the pallial PVZ, and ventral zone, corresponding to the subpallial PVZ (Adolf et al., 2007; Lindsey et al., 2007). Data of different researcher demonstrate the importance of phylogenetic studies of the pallial and subpallial proliferative zones in the forebrain of zebrafish and other fish species (Lindsey et al., 2007; Zupanc and Sîrbulescu, 2013). The presence of the NSCs involved in adult neurogenesis with a glial phenotype is demonstrated (Adolf et al., 2007; Kempermann, 2011). In particular, it is shown that the PVZ contains slowly and rapidly proliferating cells developing from adult NSCs/precursor cells with a radial glial phenotype (Adolf et al., 2007). A recent classification, based on detailed immunohistochemical labeling (März et al., 2010) and clonal analysis (Rothenaigner et al., 2011), distinguishes between three types of mitotically active cells (II, IIIa, and IIIb) and non-dividing type I cells. An examination of the pallial–subpallial zone of the zebrafish forebrain PVZ has shown that the precursor cells express various transcription and growth factors, such as Pax6 and FGF, and are putatively regulated by these factors (Adolf et al., 2007; Ganz et al., 2010). However, the ever-increasing number of studies on the forebrain pallial and subpallial PVZs still fail to sufficiently clarify certain basic aspects in this area, including anatomical boundaries of these zones, their ultrastructural composition, and specific features in their cellular and molecular organizations.

Analysis of regeneration of zebrafish telencephalon has been a subject of several recent investigations (Kishimoto et al., 2011; Kroehne et al., 2011; März et al., 2011; Baumgart et al., 2012; Kroehne and Brand, 2012). In these studies, telencephalon was chosen as the assay system because: (1) newborn neurons are not observed in the parenchyma of the telencephalic pallial subdivision under normal conditions, thus enabling an observer to distinguish between constitutive neurogenesis and regeneration, and (2) there is hope to get an insight into functional differences of teleost and mammalian telencephalic regenerative behaviour.

Since juvenile masu salmon are in the state of active growth, the constitutive production of new cells, according to our data, is rather intensive and is observed both in the brain proliferative zones and beyond them. We believe that the phylogenetic factors, demonstrating that salmonids belong to an evolutionarily ancient group, the teleosts, for which the uncompleted eversion of telencephalon and retained embryonic brain structural traits are characteristic in the postembryonic development, can also play an important role (Wullimann and Muller, 2009). A similar characteristic, referred to as *fetalization*, is typical of phylogenetically ancient vertebrate groups, the brain of which contains many morphogenetic zones with an increased proliferative potential (Artyukhin, 2008; Zupanc and Sîrbulescu, 2013). Our studies suggest that the cell proliferative potential of the medial zone is considerably higher as compared with the other zones of pallium. One more specific feature of the masu salmon medial zone is a pronounced depthward cell migration from the surface layers of pallium. The stratification of cell layers in this region has been traced in individual cases. Thus, our data suggest that the medial zone is the most important contributor to the constitutive morphogenesis of masu salmon telencephalon as compared with the remaining zones of the dorsal region.

Currently, the data on the embryonic origin, associated with homogeneity of the NSCs in the adult fish brain, are scarce. The studies on zebrafish demonstrate that the proliferative zone of the definitive pallium develops from two separate subtypes of embryonic precursors and comprises two NSC types involved in the adult neurogenesis (Dirian et al., 2014). The dorsomedial NSCs emerge owing to active amplification of the neurogenic radial glia in the embryonic telencephalon. The NSC population of the lateral zone is formed through constant supplementation of the pallial band from the pool of discrete neuroepithelial progenitors of the telencephalon posterior roof, activated in the postembryonic period and retained throughout the lifecycle. This dual origin of the zebrafish pallial proliferative zone implies a time-dependent construction of pallial regions as mosaic and adjacent compartments. Taking into account the results obtained from zebrafish, the data on a nonuniform distribution of the proliferating and HuCD+ cells in the masu salmon periventricular proliferative zones, and the ratio of migrating and proliferating cells in the brain parenchyma, we believe that the masu salmon pallial PVZ has a complex multicomponent mosaic structure and is most likely comprised of several formations containing NSCs.

After a mechanical injury of masu salmon pallium, we observed considerable changes in the cell proliferative activity in both the pallium proliferative zones and parenchyma. The most pronounced changes in the topography of proliferating cells were associated with the structural remodeling of the integrated proliferative, superficially located in the periventricular layer, and formation of the local zones of induced neurogenesis (Figure 1B, D). It should be noted that the periventricular zone as a rule was not verified. Formation of local cell clusters containing both PCNA+ elements and numerous PCNA- cells in *reactive neurogenic niches*, as well as vast populations of migrating cells and weakly labeled radial glial fibers, were observed (Figure 1B, D). These changes appeared after a mechanical injury, suggesting that they can be regarded as repair rearrangements.

Another specific feature in the post-traumatic morphogenesis of masu salmon telencephalon was the formation of the *secondary neurogenesis centers* (reactive parenchymatic niches) in the medial zone, which also possesses an elevated proliferative potential in norm (Figure 1F). After the injury, numerous PCNA- cells surrounded by undifferentiated immunonegative cells were observed. The distribution density in such reactive clusters considerably exceeded that in the adjacent regions, suggesting that these clusters can be regarded as integrated functional complexes with an increased proliferative activity. According to the data by Savel'ev (2001), the centers of secondary neurogenesis are formed in vertebrate brain when cells from the primary periventricular proliferative region migrate to deeper brain layers retaining their increased proliferative potential. Such cells continue proliferation even beyond the typical conditions of cell microenvironment in the proliferative zone and form the centers of secondary proliferation. In this process, a part of the cells continues proliferation, whereas the other part starts to differentiate. Such cell cluster outside the matrix zone typically has a higher density and heterogeneous cell composition and contains cells at different stages of differentiation. We observed these particular cell clusters in the medial zone of masu salmon telencephalon after injury.

The studies on zebrafish have demonstrated that an injury of telencephalon induces a rapid proliferation of neural precursor cells in the PVZ of the injured telencephalon hemisphere as compared with the intact one. The NSC distribution, detected using BrdU and neurogenin, demonstrates that these cells migrate laterally and reach the injured site via subpallium and pallium (Kishimoto et al., 2011, 2012). Recent studies of the zebrafish NSCs have revealed new aspects in the physiology of the radial glia. In particular, it is shown that the radial glia is present in the fish brain proliferative zones (Grandel et al., 2006). A part of the cells forming the radial glial population possesses the NSC properties, which determines a high level of reparative neurogenesis (Zupanc, 2001; Grandel et al., 2006). Cells of the fish radial glia are spread all over the brain regions along the ventricular lumen and enhance successful regeneration of injured brain (Tozzini et al., 2012). It is known that the regenerative potential of somatic stem cells, as a rule, decreases with age of animals; thus, the intensity of constitutive neurogenesis in juvenile masu salmon individuals of different ages (Pushchina et al., 2012) suggests a proliferative potential of NSCs at different ages.

Studies on zebrafish have shown that neurogenesis and oligodendrogenesis in the ventricular zone, olfactory bulb, and parenchyma of the telencephalon weaken with aging (Edelmann et al., 2013). According to our data, pallium in juvenile masu salmon displays a high level of constitutive neurogenesis (Pushchina et al., 2017). We have earlier demonstrated that the periventricular region of masu salmon of different age cohorts have a large content of the radial glia, marked by tyrosine hydroxylase, NADPH diaphorase, and GABA (Pushchina et al., 2012). The radial glial cell population in the periventricular zone remains unchanged to a considerable degree, but immunohistochemical analysis demonstrates that such cells in more adult animals rarer enter mitosis and, consequently, produce fewer neuroblasts. The activity of neuroblasts remains rather constant with age (Adolf et al., 2006; Grandel et al., 2006); thus, neuroblasts produce the same number of postmitotic neurons. The decrease in the level of neurogenesis physiologically correlates with the increase in the radial glia quiescence (Edelmann et al., 2013). An injury activates the radial glia and induces its proliferation. However, the radial glia is to a considerably lesser degree involved in neurogenesis in adult animals as compared with juveniles, suggesting certain irreversible changes in the radial glial cells during brain aging.

Muller and Wullimann (2003) have shown that PCNA labeling of proliferative regions reflects a neuromeric organization of the brain, which has also been confirmed by autoradiography and immunocytochemical (BrdU) assays. PCNA expression in cells retains for 24 h after the completion of mitosis, but the level of its activity decreases by 30% (Wullimann and Puelles, 1999). PCNA is expressed in mitotic cells during the entire proliferative cycle and has a prevalent nuclear localization (Waseem and Lane, 1990); however, cytoplasmic localization of PCNA is also observable in some cases (Vriz et al., 1992). Specific methodical features of PCNA as a cell proliferation marker in the studies of different neurogenesis stages in fish brain has made it possible to demonstrate that this marker at the early developmental stages in zebrafish (days 1–4 after fertilization) identifies neuroepithelial cells that have a rather short cell cycle in this period (Muller and Wullimann, 2003). Nonetheless, PCNA is a convenient cell proliferation marker when studying later postembryonic neurogenesis in fish (Candal et al., 2005), since the cell cycle at subsequent stages is considerably (manifold) longer as compared with the early neurogenesis.

Various lesion paradigms have been elaborated for studying the nervous tissue regeneration of fish brain and its functional repair after CNS injury. Many studies of telencephalon injuries focus on injuries of olfactory bulbs or the dorsal region in one of the telencephalon hemispheres (Zupanc, 2001; Zupanc and Sîrbulescu, 2013). In some cases, the neural repair was studied in the lateral and dorsolateral parts of telencephalon (Ayari et al., 2010). A better insight into the bony fish brain regeneration capacities requires the knowledge of the sources for neuron regeneration in injured brain. Genetic studies involving an injury of dorsal pallium demonstrate that most of regenerating neurons originate from radial glia-type progenitor cells, while differentiation of nonneurogenic cells plays a minor role in the brain repair (Kroehne et al., 2011). A systematic neuroanatomy examination of zebrafish allowed distinguishing of 16 constitutive neurogenic domens located along the brain axis (Adolf et al., 2006; Grandel et al., 2006; Kaslin et al., 2008). As has been shown, the adult NSCs are mainly associated with the ventricular system (Tozzini et al., 2012). The NSCs in fish telencephalon have radial glia morphology and express certain molecular markers (vimentin and nestin) characteristic of the mammalian NSCs (Ganz et al., 2010; März et al., 2010). The pial and subpallial constitutive neurogenic niches, identified in fish, are now regarded as homologs of the subgranullar and subventricular proliferative zones in mammalian brain (Adolf et al., 2006; Mueller and Willimann, 2009).

Telencephalon of zebrafish mostly recovers histologically by 30 dpl and becomes indistinguishable from a non-injured telencephalon after 1 year. Besides the rapid reduction of cell death, gliosis and inflammation, the regenerative/reactive neurogenesis commences with the initial proliferative response of radial glia in the endogenous proliferation domain that produces new neurons in the damaged hemisphere (Kroehne et al., 2011). As early as at 4 dpl, pulse-chase experiments show many newborn pallial neurons in the periventricular zone (Kroehne et al., 2011; März et al., 2011), and newborn parenchymal neurons become evident in the damaged brain from the first week post injury (Kroehne et al., 2011; Baumgart et al., 2012). The parenchymally located newborn neurons, which arise from the ventricularly located radial glial cells as determined by Cre-lox lineage tracing technology, home into the lesion site only after a very long chase time (21–90 dpl) (Kroehne et al., 2011). Pallial radial glia are under the control of notch signalling (Ganz et al., 2010; Kishimoto et al., 2012) which can be monitored by transgene expression under the control of the her4.1 promoter. A transgenic conditional Cre-lox recombination approach has been used to genetically mark pallial radial glia and their progeny (Kroehne et al., 2011).

Thus, PCNA immunolabeling of masu salmon pallium demonstrates the presence of the superficially located proliferative layer and individual proliferating cells with spatially specific number and localization (Pushchina et al., 2017). In general, the distribution pattern of PCNA+ cells in the dorsal telencephalon matches the corresponding patterns of other fish species; it should be noted that an important specific feature consists in organization of the medial zone, the main characteristic of which is an increased proliferative potential. Taking into account the data on the incomplete eversion of salmonid telencephalon, we tend to believe that the medial part of telencephalon is the particular region that houses the most of cell proliferation and migration events that are decisive in telencephalon formation during persistent neurogenesis. The zones of induced neurogenesis appear exclusively in the postinjury process and, in our view, are related with the brain repair.

HuCD Localization in the Proliferative Zones and Parenchyma of the Masu Salmon Pallial PVZ in Norm and after a Mechanical Injury

Most researchers regard the cells of the PCNA- brain regions as stimulated to differentiation (Candal et al., 2005) because of the presence of neural determination markers, such as Pax6, Zash-1a, Zash-1b, neurogenin, and neuroD, and neural differentiation marker HuCD (Mueller and Willimann, 2009). The newborn neurons in zebrafish pallium also expressed more mature region-specific neuronal markers such as parvalbumin and Prox1 in an appropriate manner and appeared integrated into the neural circuitry as visualized by dendritic (MAP2a+b) and synaptic markers (SV2, mGlu2). The results of HuCD labeling of pallium of juvenile masu salmon confirm these data. A comparison of the distributions of HuCD+ neurons in the dorsal, lateral, and medial zones of masu salmon pallium demonstrates that the definitive cells with a mature neuron-like phenotype, most likely involved in formation of neuronal networks, display different degrees of HuCD immunostaining. The results obtained from juvenile masu salmon differ from the data on HuCD immunolocalization after a mechanical eye injury in adult trout (Pushchina et al., 2016). The trout pallial proliferative zone also displayed a high level of HuCD activity as compared with the definitive neurons in the dorsal regions (Pushchina et al., 2016). However, neither typical neurogenic niches nor structural changes in the dorsal pallial zone, observed after an injury of pallium in juvenile masu salmon, were detectable after an optic nerve injury in adult trout. Thus, we believe that the region of brain injury, as well as the animal age, influences the pattern of postinjury repair. In particular, an injury of trout eye and optic nerve activates neurogenic niches in the optic tectum and cerebellum, but no typical neurogenic niches were detected in telencephalon (Pushchina et al., 2016). The control masu salmon juveniles displayed rather intensive constitutive neurogenesis in different parts of the pallial proliferative zone. The superficially located HuCD+ cells either formed clusters or, in some cases, even a superficial cell layer. The distribution density of immunopositive cells in the lateral zone was higher as compared with the dorsal zone; however, the most superficial layer was formed of negative cells. A high density of both HuCD+ and HuCD– cells and a moderate density of cell distribution in the proliferative region were characteristic of the medial zone.

The densitometric data for the postinjury period make it possible to distinguish between two HuCD activity levels in masu salmon pallium: intensive and moderate (Pushchina et al., 2017). Intensively labeled definitive neurons were prevalent in all the regions of the control animals. The HuCD+ neurons were surrounded by undifferentiated immunonegative cells varying in number in different zones. According to the classification by Edelmann et al. (2013), we referred these cells as to the cell population stimulated to neural differentiation but not yet expressing neural determination markers. Our studies allowed us to distinguish four types of HuCD+ cells differing in their morphology and density of HuCD immunolabeling. We believe that the differences in densitometric characteristics of HuCD+ cells may reflect the differences in the distribution pattern of this protein in the neuron cytoplasm and correspond to different degrees of neuronal differentiation of the cells formed during reparative neurogenesis. Nonetheless, we should emphasize that masu salmon brain contains neurons displaying different degrees of HuCD immunolabeling irrespective of their differentiation stage and also the described types of neurons. We also assume that different densitometric characteristics of HuCD+ cells may result from methodological aspects, in particular, the ability of antibodies to penetrate into different depths in brain sections. However, the most apparent and grounded assumption explaining the differences in HuCD labeling intensity during reparative neurogenesis, in our view, is that the produced cells are at different stages of neuronal differentiation.

The expression of HuCD protein in the pallial proliferative zones of masu salmon yearlings is rather pronounced, which agrees with the data on zebrafish (Mueller et al., 2011). The cells newly formed in dorsal telencephalon of zebrafish after an injury start to express HuCD on days 3–4 post-injury (Ayari et al., 2010; Kishimoto et al., 2011). The newly formed neurons appear both near the injury site and in the regions distant from this site. As is shown, a small population of zebrafish cells, formed within 2 days before the injury, is involved in the repair of the injured brain region (Kroehne et al., 2011). The number of HuCD expressing cells gradually increases in the subsequent days, especially in the injured zone. An analysis of the spatiotemporal pattern of HuCD+ cell distribution in zebrafish demonstrates that the newly formed cells acquire the phenotype characteristic of the neurons during migration to the injured zone (Lam et al., 2009; Edelmann et al., 2013). These data suggest that the interactions between proliferating cells in an intact fish brain also change after the injury.

Both intensively and moderately HuCD immunolabeled cells were observed in the deep layers of the masu salmon dorsal zone on day 3 post-injury, which considerably differs from the immunolabeling pattern in the control (Figure 2A, B). In our opinion, a decrease in the HuCD immunolabeling intensity in most of definitive neurons against the background of retained intensive labeling in 25% of the cells is a manifestation of the adaptive response to a mechanical injury of pallium. A redistribution of the HuCD+ cells, forming the layer of irregular clusters with retention of the immunonegative surface layer in control animals, was observed in the proliferative region of the dorsal zone.

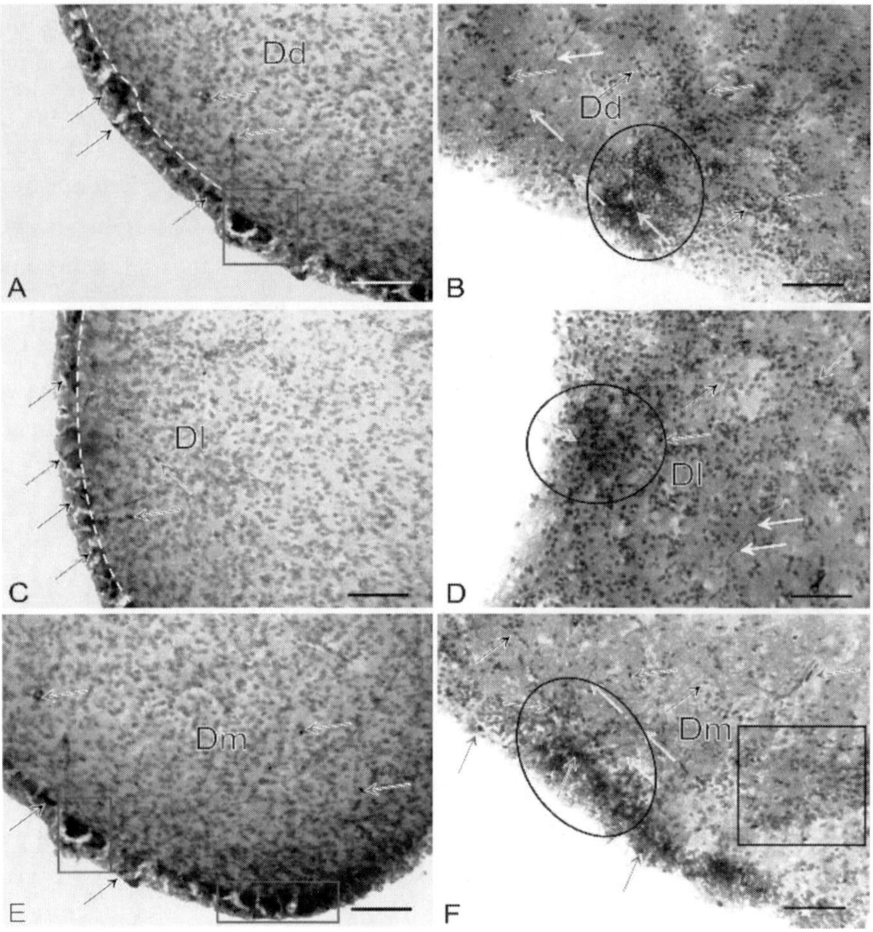

Figure 2. Distribution of the HuCD neural protein in the pallial proliferative zone and deep layers of the dorsal region in juvenile masu salmon *Oncorhynchus masou* (A, C, E) in norm and (B, D, F) after a mechanical injury: (A) dorsal (DD), (C) lateral (DL), and (E) medial (DM) zones of masu salmon telencephalon in the control (rectangles enclose regions of the periventricular zone and arrows of different colors indicate cells of types 1–4) and (B) dorsal, (D) lateral, and (F) medial regions after a mechanical injury (red arrows indicate intensively labeled neurons; white, weakly labeled; in panel (D), black arrows indicate radial glial fibers; rectangles enclose neurogenic niches containing newly formed neurons; black arrows in panel (F) indicate intensively labeled neurons in the pallial periventricular zone; white dashed line is direction of cell migration, and red line segments show different thicknesses of neurogenic layer). Scale bar is 100 μm.

The ratio of intensively to moderately HuCD-labeled definitive neurons in the lateral zone was 1:2.5, differing from that in the dorsal zone. After injury, the proliferative zone underwent considerable changes (Figure 2C, D). Two zones of induced neurogenesis were observed as neurogenic niches of different morphologies and, possibly, different origins. After an eye injury in adult trout, local clusters of HuCD+ cells were observed in the lateral proliferative zone (Pushchina et al., 2016). Neurogenic niches in the lateral zone of juvenile masu salmon individuals were surrounded by weakly immunolabeled radial glial fibers. Formation of such structural complexes producing new neurons is a consequence of activated proliferation in NSCs induced by injury (Lam et al., 2009; Edelmann et al., 2013). NSCs are able to produce neurons and glia both *in vivo* and *in vitro*, although the neuronal or glial specialization is determined by the conditions of cell microenvironment and other dynamic factors (Adolf et al., 2006; März et al., 2010; Kishimoto et al., 2011). Similar to mammals, the precursor cells in zebrafish give rise mainly to neurons rather than glial cells (Hinsch and Zupanc, 2006; Mueller et al., 2011). The causes of this phenomenon in zebrafish are still unclear; however, a limited production of astrocytes in adult brain correlates with fits their total low content in brain of bony fish.

According to Kishimoto et al. (2012), the NSCs that emerge after an injury of zebrafish telencephalon differentiate into mature neurons during 1 week post-injury. Most of these cells express Tbr1, suggesting a normal adaptation of the neurons in this region, Thus, the ventricular zones of zebrafish telencephalon are involved in the repair of nervous tissue after a telencephalon injury in adult fish.

A considerable increase in the number of intensively labeled HuCD+ cells, localized to the proliferative zone and forming migratory flows of cells with neuronal specialization, was observed in the medial zone of masu salmon pallium (Figure 2E, F). This overproduction of neurons in the medial zone after injury generally agrees with the increased proliferative potential of this zone, which we revealed by PCNA labeling. As for adult trout, we observed an increased density of HuCD+ cells in the dorsal zone after an eye injury (Pushchina et al., 2016). We believe that the increased production of neurons during the postinjury period in different telencephalon zones is a species-specific feature of salmonids and may possibly reflect certain specific phylogenetic features of the repair process characteristic of this fish group.

CONCLUSION

Stab injury experiments were conducted by inflicting a damage to telencephalon with a needle. This type of traumatic brain injury initiates a cascade of events. An immediate response to the stab is the formation of an oedema and a wave of neuronal cell death, both of which rapidly recede at 7 dpl and 3 dpl, respectively. Another reaction to a stab injury is the onset of gliosis and inflammation in the damaged hemisphere of masu salmon pallium. In the parenchyma of the damaged pallium, from 4 dpl on, a marked hypertrophy of GFAP-positive and vimentin-positive radial fibres is observed.

L-plastin-positive cells in pallium of adult zebrafish consist of a heterogeneous mixture of blood-derived leukocytes and resident microglia that become more numerous after a stab wound (Kroehne et al., 2011; Baumgart et al., 2012). Proliferation levels of different cell types also increase after damage to telencephalon in adult zebrafish, peaking at 3 dpl in the endogenous ventricular proliferation domain, as well as ectopically in the damaged parenchyma. In the endogenous ventricular proliferation zone, radial glial cells expressing GFAP, vimentin, nestin and glutamine synthetase specifically upregulate their proliferation upon a stab injury. In the parenchyma of zebrafish telencephalon, mostly L-plastin-positive cells contribute to the proliferating population (Kroehne et al., 2011); however, a small fraction of oligodendrocyte precursors proliferate as well (Kroehne et al., 2011; März et al., 2011). Also, proliferating endothelial cells can be observed (Kroehne et al., 2011). Slightly elevated proliferation rates can still be detected at 2 weeks post-injury but decline subsequently. Likewise, the initial early inflammatory response does not persist and neither hypertrophic radial glial processes nor a surplus of L-plastin-positive cells are detected by 30 dpl. It is important that the glial scar does not form (Kroehne et al., 2011).

This is in marked contrast to the reaction of mammalian telencephalon to a damage where regeneration is hampered by gliosis, inflammation and subsequent formation of the glial scar that inhibits regeneration of neural tissue (Fitch and Silver, 2008; Sofroniew, 2009).

The study of proliferative activity of pallium in juvenile masu salmon *O. masou* has allowed us to identify the superficially located periventricular proliferative zone, which corresponds to the pallial periventricular zone (PVZ) of other fish species, including the dorsal, lateral, and medial compartments. A superficially located periventricular proliferative area with PCNA+ cells, which corresponds to the PVZ of other fish species, including its dorsal, lateral, and medial compartments, has been discovered in pallium of juvenile masu salmon. The PCNA+ cells are also identified in the parenchyma of intact masu salmon brain, and their maximum concentration is observed in the medial zone. In intact brain, solitary proliferating cells were identified in parenchymal regions with their maximum concentration in the medial zone. After a mechanical injury, the zones of reparative neurogenesis — neuroepitelial neurogenic niches and zones of reactive neurogenesis surrounded by radial glial fibers — appear in masu salmon pallium. An increase in the proliferative activity is characteristic of all compartments of the PVZ (especially of the medial zone), including its deep parenchymal layers. The PVZ of juvenile masu salmon contains clusters of undifferentiated HuCD+ neurons. A change in the HuCD+ cell topography is observed in a mechanically injured masu salmon pallium: reactive neurogenic niches in the lateral zone and an increase in the cell distribution density and cell migration patterns in the medial zone. A high level of constitutive neurogenesis is characteristic of juvenile masu salmon brain. Undifferentiated HuCD+ neurons form clusters in the PVZ of intact brain in juvenile masu salmon. Neurons at different stages of maturation, differing in the level of HuCD activity, were detected in the parenchymal layers of the dorsal region. A mechanical injury of pallium in masu salmon is accompanied by a change in the topography of HuCD+ cells. Formation of reactive neurogenic niches is characteristic of the lateral zone, while the medial zone displays an increase in the density of immunopositive and negative cells, as well as a distinct pattern of cell migration. HuCD and

PCNA immunolabeling patterns in intact pallium suggest the intensive constitutive neurogenesis in the pallial PVZ of juvenile masu salmon.

The colocalization of newborn cells with the neuronal marker HuCD showed that her4.1-positive radial glia generates neurons. A two-third of the newborn neurons were located in a periventricular position close to the pallial proliferation zone that corresponds to their regular target area. A one-third of all HuCD-positive newborn neurons were found in a parenchymal position that was usually not attained by newborn neurons neither in undamaged brains nor in the control hemisphere.

Thus, the damage site specifically attracts newborn neurons from the endogenous proliferation domain that are rerouted to populate and integrate into this site (Kroehne et al., 2011). It will be important to decode the signals that emanate from the damage site, which act to trigger the regenerative response. Chemokines and their receptors have been suggested as possible candidates. They have been reported to localize to radial glia in telencephalon (Diotel et al., 2010), where their overexpression enhances the proliferative response after an injury (Kizil et al., 2012). Other, still unidentified, pathways may also funnel into a regenerative response to trigger there generative pathway at different angles (Kyritsis et al., 2012). This observation confirms the long-lasting suggestion (Kirsche and Kirsche, 1961) that periventricular progenitors of the "matrix zones" produce neuronal progeny, which are engaged in regeneration.

Funding

This work was supported by a grant from the President of the Russian Federation (No. MD-4318.2015.4), a grant from the Program for Basic Research of the Far East Branch of the Russian Academy of Sciences 2015–2017 (No. 15-I-6-116, section III).

REFERENCES

Adolf, B., Chapouton, P., Lam, C.S., Topp, S., Tannhauser, B., Strahle, U., Gotz, M., and Bally-Cuif, L., Conserved and acquired features of adult neurogenesis in the zebrafish telencephalon, *Dev. Biol.,* 2006, vol. 295, pp. 278–293.

Artyukhin, E.N., *Osetrovye (Ecologiya, geograficheskoe rasprostranenie i filogeniya) (Acipenseridae: Ecology, Geographic Distribution, and Phylogeny),* St. Petersburg: Izd. SPbGU, 2008.

Ayari, B., Hachimi, K.H., Yanicostas, C., Landoulsi, A., and Soussi-Yanicostas, N., Prokineticin 2 expression is associated with neural repair of injured adult zebrafish telencephalon, *J. Neurotrauma,* 2010, vol. 27, pp. 959–972.

Baumgart, E.V., Barbosa J.S., Bally-Cuif L., Gotz M., and Ninkovic J., Stab wound injury of the zebrafish telencephalon: A model for comparative analysis of reactive gliosis, *Glia,* 2012, vol. 60, pp. 343–357.

Berninger, B., Hack, M.A., and Gotz, M., Neural stem cells: on where they hide, in which disguise, and how we may lure them out, *Handb. Exp. Pharmacol.,* 2006, vol. 174, pp. 319–360.

Candal, E., Anadon, R., Bourrat, F., and Rodriguez-Moldes, I., Cell proliferation in the developing and adult hindbrain and midbrain of trout and medaka (teleosts): a segmental approach, *Brain Res. Dev. Brain Res.,* 2005, vol. 160, pp. 157–175.

Chojnacki, A.K., Mak, G.K., and Weiss, S., Identity crisis for adult periventricular neural stem cells: subventricular zone astrocytes, ependymal cells or both?, *Nat. Rev. Neurosci.,* 2009, vol. 10, pp. 153–163.

Diotel, N., Vaillant, C., Gueguen, M.M., Mironov, S., Anglade I., Servili A., Pellegrini, E., and Kah, O., Cxcr4 and Cxcl12 expression in radial glial cells of the brain of adult zebrafish, *J. Comp. Neurol.,* 2010, vol. 518, pp. 4855–4876.

Diotel, N., Vaillant, C., Gabbero C., Mironov, S., Fostier, A., Gueguen, M. M., Anglade, I., Kah, O., and Pellegrini, E., Effects of estradiol in adult neurogenesis and brain repair in zebrafish, *Horm. Behav.,* 2013, vol. 63, pp.193–207.

Dirian, L., Galant, S., Coolen, M., Chen, W., Bedu, S., Houart, C., Bally-Cuif, L., and Foucher, I., Spatial regionalization and heterochrony in the formation of adult pallial neural stem cells, *Dev. Cell,* 2014, vol. 30, pp. 123–136.

Edelmann, K., Glashauser, L., Sprungala, S., Hesl, B., Fritschle, M., Ninkovic, J., Godinho, L., and Chapouton, P., Increased radial glia quiescence, decreased reactivation upon injury and unaltered neuroblast behavior underlie decreased neurogenesis in the aging zebrafish telencephalon, *J. Comp. Neurol.,* 2013, vol. 521, pp. 3099–3115.

Fitch, M.T., and Silver J., CNS injury, glial scars, and inflammation: inhibitory extracellular matrices and regeneration failure, *Exp. Neurol.,* 2008, vol. 209, pp. 294–301.

Ganz, J., Kaslin, J., and Hochmann, S., Heterogeneity and Fgf dependence of adult neural progenitors in the zebrafish telencephalon, *Glia,* 2010, vol. 58, pp. 1345–1363.

Grandel, H., Kaslin, J., Ganz, J., Wenzel, I., and Brand, M., Neural stem cells and neurogenesis in the adult zebrafish brain: origin, proliferation dynamics, migration and cell fate, *Dev. Biol.,* 2006, vol. 295, pp. 263–277.

Hinsch, K., and Zupanc, G.K., Isolation, cultivation, and differentiation of neural stem cells from adult fish brain, *J. Neurosci. Methods,* 2006, vol. 158, pp. 75–88.

Hutton, C., Dery, N., and Rosa, E., Mitigating the effects of stress on the hippocampus with diet and exercise, in *Abstr. Intern. Confer. Adult Neurogenesis: Evolution, Regulation and Function,* Dresden, Germany, 2015, Abstr. 23.

Kaneko, N., and Sawamoto, K., Adult neurogenesis and its alteration under pathological conditions, *Neurosci. Res.,* 2009, vol. 63, pp. 155–164.

Kaslin, J., Ganz, J., and Brand, M., Proliferation, neurogenesis and regeneration in the non-mammalian vertebrate brain, *Philos. Trans. R. Soc. Lond. B. Biol. Sci.,* 2008, vol. 363, pp. 101–122.

Kempermann, G., *Adult Neurogenesis 2 - Stem Cells and Neuronal Development in the Adult Brain,* New York: Oxford Univ. Press, 2011.

Kempermann, G., *Adult neurogenesis, in Neuroscience in the 21st Century,* Pfaff, D.W., Ed., New York, 2013, pp. 161–178.

Kishimoto, N., Alfaro-Cervelloc, C., Shimizu, K., Asakawa, K., Urasaki, A., Nonaka, S., Kawakami, K., Garcia-Verdugo, J.M., and Sawamoto, K., Migration of neuronal precursors from the telencephalic ventricular zone into the olfactory bulb in adult zebrafish, *J. Comp. Neurol.,* 2011, vol. 519, pp. 3549–3565.

Kirsche, W., and Kirsche K. Experimental studies on the problem of regeneration and function of the tectum opticum of *Carassium carassium* L., *Z. Mikrosk. Anat. Forsch.,* 1961, vol. 67, pp. 140–182.

Kishimoto, N., Shimizu, K., and Sawamoto, K., Neuronal regeneration in a zebrafish model of adult brain injury, *Dis. Model. Mech.,* 2012, vol. 5, pp. 200–209.

Kizil, C., Dudczig S., Kyritsis N., Machate A., Blaesche J., Kroehne V., and Brand M., The chemokine receptor cxcr5 regulates the regenerative neurogenesis response in the adult zebrafish brain, *Neural. Dev.,* 2012, vol. 7:27, doi: 10.1186/1749-8104-7-27.

Kroehne, V., Freudenreich, D., Hans, S., Kaslin, J., and Brand, M., Regeneration of the adult zebrafish brain from neurogenic radial glia-type progenitors, *Development,* 2011, vol. 138, pp. 4831–4841.

Kyritsis, N., Kizil, C., Zocher, S., Kroehne, V., Kaslin, J., Freudenreich, D., Iltzsche, A., and Brand, M., Acute inflammation initiates the regenerative response in the adult zebrafish brain, *Science,* 2012, vol. 338, pp. 1353-1356.

Lam, C.S., Marz, M., and Strahle, U., GFAP and nestin reporter lines reveal characteristics of neural progenitors in the adult zebrafish brain, *Dev. Dyn.,* 2009, vol. 238, pp. 475–486.

Lema, S.C., Hodges, M.J., Marchetti, M.P., and Nevitt, G.A., Proliferation zones in the salmon telencephalon and evidence for environmental influence on proliferation rate, *Comp. Biochem. Physiol.,* 2005, vol. 141A, pp. 327–335.

Lillesaar, C., Stigloher, C., Tannhauser B., Wullimann, M.F., and Bally-Cuif L., Axonal projections originating from raphe serotonergic neurons in the developing and adult zebrafish, *Danio rerio*, using transgenics to visualize raphespecific pet1 expression, *J. Comp. Neurol.,* 2009, vol. 512, pp. 158–182.

Lindsey, B.W., Darabie, A., and Tropepe, V., The cellular composition of neurogenic periventricular zones in the adult zebrafish forebrain, *J. Comp. Neurol.,* 2007, vol. 520, pp. 2275–2316.

Lindsey, B.W., Donato, S., Kaslin, J., and Tropepe, V., Sensory-specific modulation of adult neurogenesis in sensory structures is associated with the type of stem cell present in the neurogenic niche of the zebrafish brain, *Eur. J. Neurosci.,* 2014, vol. 40, pp. 3591–3607.

Lindsey, B.W., and Tropepe, V., Changes in the social environment induce neurogenic plasticity predominantly in niches residing in sensory structures of the zebrafish brain independently of cortisol levels, *Dev. Neurobiol.,* 2014, vol. 74, pp. 1053–1077.

März, M., Chapouton, P., Diotel, N., Vaillant, C., Hesl, B., Takamiya, M., Lam, C.S., Kah, O., Bally-Cuif, L., and Strahle, U., Heterogeneity in progenitor cell subtypes in the ventricular zone of the zebrafish adult telencephalon, *Glia,* 2010, vol. 58, pp. 870–888.

März, M., Schmidt, R., Rastegar, S., and Strahle, U., Regenerative response following stab injury in the adult zebrafish telencephalon, *Dev. Dyn.,* 2011, vol. 240, pp. 2221–2231.

Menuet, A., Pellegrini. E., Brion F., Gueguen, M.M., Anglade, I., Pakdel, F., and Kah, O., Expression and estrogen-dependent regulation of the zebrafish brain aromatase gene, *J. Comp. Neurol*

Mouriec, K., Gueguen, M.M., Manuel, C., Percevault, F., Thieulant, M.L., Pakdel, F., and Kah, O., Androgens upregulate cyp19a1b (aromatase B) gene expression in the brain of zebrafish (*Danio rerio*) through estrogen receptors, *Biol. Reprod.,* 2009, vol. 8, pp.889–896.

Mueller, T., and Wullimann, M.F., An evolutionary interpretation of teleostean forebrain anatomy, *Brain Behav. Evol.,* 2009, vol. 74, pp. 30–42.

Mueller, T., Donga, Z., Berberoglua, M.A., and Guo, S., The dorsal pallium in zebrafish, *Danio rerio* (Cyprinidae, Teleostei), *Brain Res.,* 2011, vol. 1381, pp. 95–105.

Muller, T., and Wullimann, M., Anatomy of neurogenesis in the early zebrafish brain, *Dev. Brain Res.,* 2003, vol. 140, pp. 137–155.

Perez, M.R., Pellegrini, E., Cano-Nicolau, J., Gueguen, M.M., Menouer-Le Guillou, D., Merot, Y., Vaillant, C., Somoza, G.M., and Kah, O., Relationships between radial glial progenitors and 5-HT neurons in the paraventricular organ of adult zebrafish — potential effects of serotonin on adult neurogenesis, *Eur. J. Neurosci.,* 2013, vol. 38, pp.3292–3301.

Pushchina, E.V., Fleishman, M.Yu., and Timoshin, S.S., Proliferative zones in the brain of the Amur sturgeon fry. Interactions with neuromeres and migration of secondary matrix zones, *Russ. J. Dev. Biol.,* 2007, vol. 38, pp. 286–293.

Pushchina, E.V., Obukhov, D.K., and Varaksin, A.A., Neurochemical markers of cells of the periventricular brain area in the masu salmon *Oncorhynchus masou* (Salmonidae), *Russ. J. Dev. Biol.,* 2012, vol. 43, pp. 35–48.

Pushchina, E.V., Varaksin, A.A., and Obukhov, D.K., Reparative neurogenesis in the brain and changes in the optic nerve of adult trout *Oncorhynchus mykiss* after mechanical damage of the eye, *Russ. J. Dev. Biol.,* 2016, vol. 47, pp. 11–32.

Pushchina, E.V., Zharikova, E.I., and Varaksin A.A., Persistent and reparative neurogenesis in the juvenile masu salmon *Oncorhynchus masou* telencephalon after mechanical injury, *Rus. J. Dev. Biol.*, 2017, vol. 48, pp. 307–320.

Rothenaigner, I., Krecsmarik, M., Hayes, J.A., Bahn, B., Lepier, A., Fortin, G., Gotz, M., Jagasia, R., and Bally-Cuif, L., Clonal analysis by distinct viral vectors identifies bona fide neural stem cells in the adult zebrafish telencephalon and characterizes their division properties and fate, *Development,* 2011, vol. 138, pp. 1459–1469.

Savel'ev, S.V., *Sravnitel'naya anatomiya nervnoi sistemy pozvonochnykh* (Comparative Anatomy of the Nervous System of Vertebrates), Moscow: Geotar-Med, 2001.

Sofroniew, M.V., Molecular dissection of reactive astrogliosis and glial scar formation, *Trends Neurosci.,* 2009, vol. 32, pp. 638–647.

Stocum, D.L., *Regenerative Biology and Medicine,* Acad. Press, Burlington, 2006.

Than-Trong, E., and Bally-Cuif, L., Radial glia and neural progenitors in the adult zebrafish central nervous system, *Glia,* 2015, vol. 63, pp. 1406–1428.

Tozzini, E.T., Baumgart, M., Battistoni, G., and Cellerino, A., Adult neurogenesis in the short-lived teleost *Nothobranchius furzeri*: localization of neurogenic niches, molecular characterization and effects of aging, *Aging Cell,* 2012, vol. 11, pp. 241–251.

Vriz, S., Lemaitre, J.M., and Leibovici, M., Comparative analysis of the intracellular localization of c-Myc, cFos, and replicative proteins during cell cycle progression, *Mol. Cell. Biol.,* 1992, vol. 12, pp. 3548–3555.

Waseem, N.H. and Lane, D.P., Monoclonal antibody analysis of the proliferating cell nuclear antigen (PCNA) structural conservation and the detection of a nucleolar form, *J. Cell Sci.,* 1990, vol. 96, pt. 1, pp. 121–129.

Wullimann, M.F., and Muller, T., Teleostean and mammalian forebrains contrasted: evidence from genes to behavior, *J. Comp. Neurol.,* 2004, vol. 475, pp. 143–162.

Wullimann, M. and Puelles, L., Postembrionic neural proliferation in the zebrafish forebrain and its relationship to prosomeric domains, *Anat. Embryol.,* 1999, vol. 329, pp. 329–348.

Zupanc, G.K., Adult neurogenesis and neuronal regeneration in the central nervous system of teleost fish, *Brain Behav. Evol.,* 2001, vol. 58, pp. 250–275.

Zupanc, G.K. and Sîrbulescu, R.F., Teleost fish as a model system to study successful regeneration of the central nervous system, *Curr. Top. Microbiol. Immunol.*, 2013, vol. 367, pp. 193–233.

Zupanc, G.K., Kompass, K.S., Horschke, I., Ott, R., and Schwarz, H., Apoptosis after injuries in the cerebellum of adult teleost fish, *Exp. Neurol.,* 1998, vol. 152, pp. 221–230.

In: Teleosts
Editor: Mikael Herleif
ISBN: 978-1-53613-660-9

Chapter 3

NEURAL STEM CELLS IN THE CEREBELLUM OF JUVENILE MASU SALMON (*ONCORHYNCHUS MASOU*) AFTER MECHANICAL INJURY

***E. V. Pushchina*[1,2], *M. E. Stukaneva*[1] and *A. A. Varaksin*[1]**

[1]National Scientific Center of Marine Biology, Far Eastern Branch of the Russian Academy of Sciences, Vladivostok, Russia

[2]A. A. Bogomoletz Institute of Physiology, National Academy of Sciences of Ukraine, Kiev, Ukraine

ABSTRACT

The objective of this chapter was to study proliferation processes and the role of radial glia and neural stem cells in the event of injurious action on cerebellum of masu salmon *Oncorhynchus masou* juvenile. Using the immunoperoxidase staining of the glial fibrillary acidic protein (GFAP), doublecortin (DC) and proliferating cells nuclear antigen (PCNA), processes of proliferation and gliogenesis after mechanical trauma of cerebellum of juvenile masu salmon O. masou were studied. After the trauma, the intensity of proliferation and migration processes varies in different zones. Proliferation processes decrease after the trauma in lateral and basal zones, and migration increases. In the dorsal zone, on the contrary, migration processes significantly decrease and proliferation increases. In the dorsal matrix zone of cerebellum, intense cell proliferation was detected. In the dorsal, lateral, and basal zone of the molecular layer of cerebellum after traumatic injury, neurogenic niches containing PCNA and cells, as well as a heterogeneous population of PCNA- cells, were identified. At the location of neurogenic niches, fibers of radial glia and small single, intensely or moderately labeled GFAP cells were discovered. As a result of injury, GFAP+ fibers of radial glia, which form differently directed, radially oriented bundles, appeared in the dorsal matrix zone. Such structural formations were not discovered in intact animals. We suppose that, after the trauma, a structural reconstruction, connected with partial spatial reorientation of the radial glia fibers, and formation of specific directions for cells formed in this zone occur in the dorsal matrix zone. As a result of the trauma, elements of the radial glia, including both the cells possessing typical morphology and the cell fragments represented by long radially oriented processes or cell body, containing initial fragments of radial fibers, appeared in masu salmon's cerebellum.

INTRODUCTION

Fish are quite an interesting subject for investigation of neurogenetic processes, since, during the period of their postembryonic development, the formation of brain, growth, and differentiation of integrative centers are very active (Zupanc, 2006). The peculiarity of adult neurogenesis in teleost fish is the ability to form large amounts of new cells. Studies on the fresh-water fish, *Danio rerio* and *Apteronotus leptorhynchus*, have shown that proliferation rate in these species is by 1–2 orders of magnitude higher than in mammals (Zupanc and Zupanc, 2006). A mass increase of intensity of cell proliferation is observed in bony fish as a response to brain damage. Complete recovery of neural tissue occurs due to colonization of the injured area with new neurons. In mammals, the long persistency of newborn cells, as a rule, is weakened, and their brain is restricted in its ability to recover (Zupanc and Sîrbulescu, 2011).

Cerebellar morphogenesis of zebrafish has been repeatedly investigated in recent years with a focus on progenitor migration during embryonic and early larval stages (Chaplin et al., 2010; Wullimann et al., 2011). However, since adult teleosts add large numbers of new cells to the cerebellum (Zupanc and Horschke, 1995; Grandel et al., 2006; Hinsch and Zupanc, 2006; Zupanc and Sirbulescu, 2011), it is clear that teleost cerebellar morphogenesis continues after embryogenesis. Recently, Kaslin et al. (2009) have studied the continued generation of new granule neurons during adult stages with the aim to determine the identity of progenitor/stem cells and their progeny.

The basis of this is postnatal neurogenesis, the process beginning with division of precursor cells, their subsequent migration, differentiation of newly formed cells and ending with integration of new neurons in the neural network (Kempermann, 2008). The phenomenological manifestations of neurogenesis in adult fish are similar to those in embryonic brain of birds and mammals (Levine et al., 1994). It is supposed that adult neurogenesis plays an important role in quantitative harmonization and coordination of CNS neurons in motoric brain structures. In mammals, body growth occurs due to the increase in size of muscle fibers rather than in their number. Number of muscle fibers increases throughout the lifecycle of fishes, and this probably requires corresponding increase in neuron units to control the muscle activity and for the quantitative harmonization of central neurons and peripheral sensor elements. In bony fishes, production of new neurons in association with a sufficiently long survival of their significant fraction leads to continuous growth of brain and its structures. In contrast to mammals, where central parts of sensory systems become completely formed and equivalent to the quantity of sensory receptors by the moment of birth and/or shortly after it, the growth and development of sensory projections in fish brain can continue throughout the lifecycle because of the necessity of adaptation to continuous significant increase in body size and amount of incoming sensory information. The continuous formation of new neurons in adult fish is provided by the presence of large amounts of cerebral radial glia (Onteniente et al., 1983; Pellegrini et al., 2007), which are known to perform a function of neural stem cells during embryonic neurogenesis in mammals (Noctor et al., 2002; Kriegstein and Alvarez-Buylla, 2009). In mammals, new neurons form mainly in the following two zones of brain: (1) in the frontal part of the subventricular zone of the lateral ventricle, from which immature neurons migrate in the composition of the so-called “rostral migratory stream” to olfactory bulbs where they differentiate into granular neurons and periglomerular interneurons (Curtis et al., 2007), and (2) in the area located in the subgranular zone of the dentate gyrus from which new cells migrate for short distances to the granular layer of hippocampal cells and develop into mature granular neurons (Seri et al., 2004). In mouse brain after inflicting mechanical damage, proliferation of resting progenitor cells in cortex parenchyma was discovered (Ahmed et al., 2012). There is an increase in numbers of endogenic stem cells forming neurospheres when grown *in vitro*. Secondary spheres in case of differentiation yield three types of neural stem cells (NSC), which indicate their multipotency in the cortex and hippocampus alike (Ahmed et al., 2012).

Mature bony fish display high neurogenic potential in many cerebral areas. Zones of initial proliferation in fish are located along the brain ventricle and form 16 accumulations along the rostro-caudal axis of the brain (Ekström et al., 2001). The brain areas, where quite intense cell division takes place in adult individuals, are considered as zones of secondary neurogenesis being localized in telencephalon, optical tectum, and cerebellum, as well as in thalamus and brain axis (Zupanc and Ott, 1999).

Fish are able to regenerate neural tissue after a mechanic or chemical damage in cerebellum, telencephalon, olfactory bulbs, and retina, and this is supposedly closely allied to the processes of adult neurogenesis (Takeda et al., 2008; Kroehne et al., 2011; März et al., 2011; Diotel et al., 2013; Zupanc and Sîrbulescu, 2013). It has been proven than the neural tissue of fish regenerates after damage within a few weeks (Zupanc et al., 2003). Replenishment of a large number of cells lost as a result of damage occurs from different sources, such as radial glia, centers of initial and secondary proliferation, and neurogenic niches (Zupanc, 2011).

Evolutionary ancient groups like salmon fish (Salmonidae) are characterized by high concentration of undifferentiated elements in the matrix zones of brain and parenchyma alike (Pouwels, 1978). Cerebellum is a generally accepted brain zone for the study of morphogenetic processes, since it increases the proliferative activity both under normal conditions and in case of traumatic impact (Pouwels, 1978; Zupanc and Sîrbulescu, 2013). Among poikilotherms, cerebellum retains the embryonic properties only in mature bony fish, which is also typical for other areas of their brain (Margotta et al., 2004).

Recently, it has been established that the glial fibrillary acidic protein (GFAP) is one of the markers of NSC (Kriegstein and Alvarez-Buylla, 2009; Ahmed et al., 2012). The proliferating cell nuclear antigen (PCNA) is used in the study of proliferative cell populations (Candal et al., 2005). The objectives of the chapter were to study the proliferation processes and the role of the glia and neural stem cells in the event of damage to cerebellum of juvenile masu salmon *Oncorhynchus masou*.

Comparative Aspects of Adult Neural Stem Cell Activity in Amphibians and Fish

During the time of Altman's initial studies of rodents, reports on proliferation zones in brain of adult amphibians and fish were also published (Kirsche, 1967). In Amphibia, proliferation and neurogenesis are more widespread than in amniotes and are localized in telencephalon, preoptic region, hypothalamus, thalamus, midbrain and cerebellum in urodeles and anurans (Margotta et al., 2004; Raucci et al., 2006). Unfortunately, there are no many reports on constitutive neurogenesis in amphibians, but a recent study confirms constitutive neurogenesis in the forebrain and midbrain of the adult bullfrog *Rana catesbeiana* discovered by pulse-chase experiments using co-labelling of newborn cells with a marker of early neurogenesis (Simmons et al., 2008). On the contrary, the aquatic salamander *Notophthalmus viridescens* does not show constitutive neurogenesis in the midbrain. With a specific lesion of midbrain dopaminergic neurons, however, quiescent ependymoglial cells can be activated in this species to proliferate and regenerate the lost cells (Berg et al., 2010).

In teleosts, the initial reports on proliferation zones in ventricular regions of the adult brain were subsequently extended to other teleost species and uncovered the presence of constitutive proliferation zones/neurogenic zones all along the rostro-caudal axis of the brain (Kuroyanagi et al., 2010; Tozzini et al., 2011). The widespread neurogenesis in both anamniote groups has been attributed to the protracted, perhaps lifelong, growth of the body and, notably, sensory organs. In both groups, freshly hatched animals are considerably smaller than adults. Thus, the constitutive neurogenesis observed in growing teleosts and amphibians may be attributable to a growing demand of CNS processing power in conjunction with increasing sensory input (Kaslin et al., 2008). This is nicely exemplified by the growing retina and optic tectum in amphibians and fish (Cerveny et al., 2012) and the increasing number of taste buds and growth of the chemosensory lobes in medulla in cyprinids (Komada, 1994).

It is important to note that brain proliferation zones in adult zebrafish reflect a more general proliferation pattern along the ventricular walls of teleost brain (Grandel and Brand, 2013). The teleost proliferation zones are not ill-defined but distinctly localized in all the subdivisions of brain along the rostro-caudal axis. From 12 to 16 distinct proliferation zones have been recognized as distantly related between teleost species such as stickleback *Gasterosteus aculeatus*, brown ghost *Apteronotus leptorhynchus*, and zebrafish *Danio rerio*. With the exception of minor differences, the discrepancy in number reflects the investigators' interpretation rather than an actual lack or increase of zones in these species (Zupanc and Horschke, 1995; Ekström et al., 2001; Zupanc, 2006). The partially characterized brain proliferation zones of medaka *Oryzias latipes* and *Nothobranchius furzeri* likewise correspond to the general proliferation pattern; however, observations on the posterior hindbrain in these species are missing (Kuroyanagi et al., 2010; Tozzini et al., 2011). Thus, zebrafish suits well to study adult teleost brain proliferation, especially with the opportunity to elucidate the organization of proliferation zones on a cellular level, the contribution of newborn cells to normal brain function, the fate of restriction of progenitor/stem cells and, hence, their potential to repair brain parts after an injury of brain tissue.

PROLIFERATION AND MIGRATION OF CELLS IN CEREBELLUM OF FISH

The largest formation containing PCNA with cells was a dorsal matrix zone of *O. masou* cerebellum. It is located in the dorsomedial part of cerebellum's body, at the boundary between the molecular and granular layers, and the probability density of PCNA with cells in it was higher compared to the adjoining areas of the molecular layer. According to previously conducted investigations, in the composition of this zone, we identified four types of cells in a normal state and after the damage (Stukaneva et al., 2015).

In the lateral and basal zones, proliferation processes noticeably decrease after the damage and migration processes increase (Figure 1A, B). We explain it by the fact that newly formed cells begin to radially and tangentially migrate in large amounts within two days after the damage, and this process can intensify in the following days. Within the boundaries of the lateral zone, we discovered dense clusters of intensely labeled PCNA cells separated from each other by immune- negative zones. Such clusters were identified as neurogenic niches (Figure 1C). After mechanical injury, proliferation and migration processes change their activity in different zones (Figure 1D).

Thus, results of PCNA immune labeling indicate that the proliferative activity in *O. masou* is typical mostly for the dorsal matrix area of cerebellum that agrees with the data on other fish, such as *D. rerio* (Zupanc, 2006) and *Apteronotus leptorhynchus* (Sîrbulescu et al., 2015). The results of our observations allow us to assume that the structural changes detected have a direct relation to processes of reparative neurogenesis. Thus, the obtained results are a demonstration that, after the damage, the main volume of cell proliferation is found in the cerebellar dorsal matrix zone.

Short-term bromodeoxyuridine (BrdU) labeling experiments or PCNA staining showed widespread proliferation in the cerebellar molecular layer of zebrafish. Newborn cells migrate radially along vimentin/ GFAP/BLBP-expressing radial fibres towards the surface of the molecular layer, where they migrate tangentially and disperse laterally to finally dive in a radial direction through the cerebellar molecular layer into the granular layer, where they differentiate into granule neurons (Grandel and Brand, 2013). Migrating progenitors and differentiated cerebellar granule neurons express the transcription factors Zic1, Zic3, Pax3 and Neurod (Kaslin et al., 2009). Migration through the molecular layer occurs rapidly. Within 6 days, the molecular layer of zebrafish cerebellum becomes devoid of labeled cells as they reach the granular layer (Grandel and Brand, 2013). This behavior is also observed with most cells above the dorsal matrix zone, except for label-retaining cells that remained after up to 6 weeks of chase and still showed the initial S-phase marker plus either pH3 or PCNA immunoreactivity, indicating that they were still proliferating. Such cells are considered the presumptive stem cells in other systems (Doetsch et al., 1999). Thus, Kaslin et al. (2009) determined their molecular character-ristics. Tg (nestin-GFP) animals showed transgene expression in most proliferating cells above the granular layer “cap”. Tg (nestin-GFP) was also found to co-localize with label retaining cells at this position. In addition, these cells maintained expression of Sox2, Meis and Musashi 1 that were shown to localize to stem or progenitor cells in other systems (Morales and Hatten 2006). Tg (nestin-GFP)-expressing progenitors also were distinctly polarized, displaying apical adherence junctions and expressing apical markers such as ZO1, β-catenin, γ-tubulin and aPKC (Grandel and Brand, 2013). As in the case of the subpallial proliferation domain and in the periventricular margin of optic tectum, the cerebellar progenitors displayed a neuroepithelial phenotype.

To underscore this finding, Tg (nestin-GFP)-positive cells did not co-stain with the radial glia markers vimentin/GFAP/BLBP. Radial glia in the cerebellar midline consists of a non-proliferating population and may

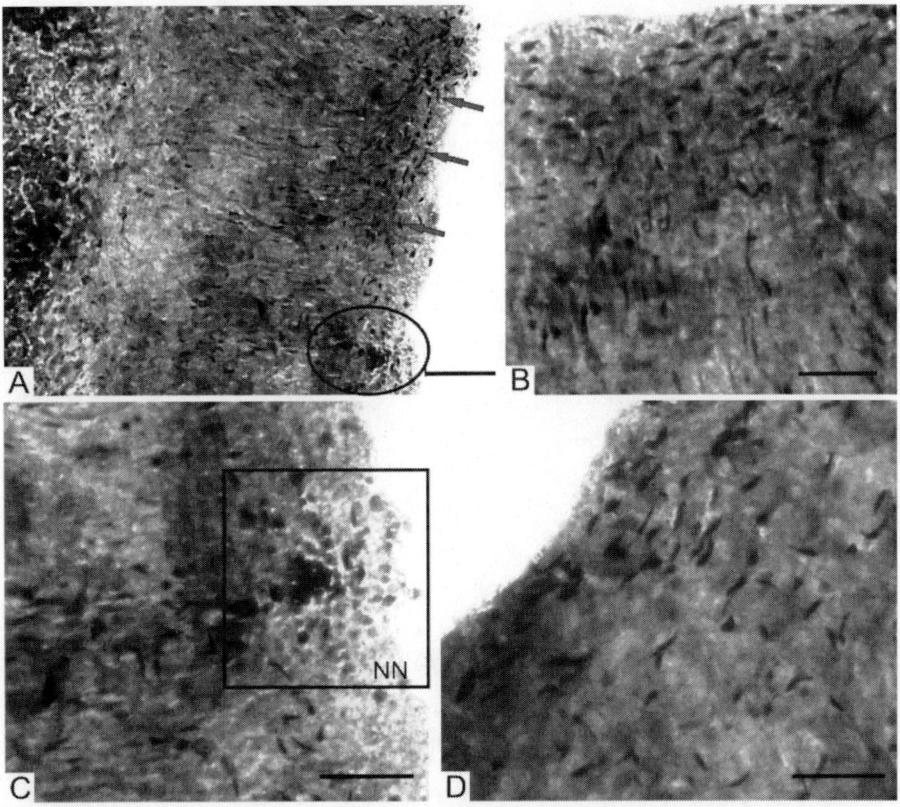

Figure 1. Localization of proliferative cell nuclear antigen (PCNA) in cerebellum of *Oncorhynchus masou* 2 days after stab-wound injury of cerebellum: (A) patterns of tangential migration (red arrows) and neurogenic niche (in oval) in the molecular layer of cerebellum; (B) radial migration of PCNA+ cells; (C) neurogenic niche (in square); (D) tangential migration of PCNA+cells. Scale bar: A, 100 μm; B–D, 50 μm.

serve to guide the migrating progeny of the neuroepithelial progenitor population. How do neuroepithelial granule cell progenitors become situated within *corpus cerebelli*, apparently far off from the ventricular system? Kaslin et al. (2009) studied the development of the cerebellar stem cell niche during embryonic and larval stages which showed that cerebellar progenitors were derived from the upper rhombic lip as in rodents and birds.

Within the first 10 days, the rate of cell proliferation in the area of brain trauma in bony fish increases several times compared to other areas of cerebellar body. Experiments with labeling by BrdU have shown that cells formed 2 days before the trauma take part in the regeneration process (Zupanc and Ott, 1999). This observation presupposes a presence of a link between continuous cell proliferation in an intact brain and restoration of the area damaged due to injury. A retrograde tracing in combination with BrdU labeling of S-phase have shown that new granular neurons are projected into the molecular layer of cerebellum (Zupanc and Ott, 1999). This fact allows us to assume that these neurons integrate into the already existing neuron network of cerebellum.

In zebrafish, however, a distinct morphogenetic process during embryonic and early larval stages (24 h to 7 days) displaces medially located rhombic lip progenitors and a small portion of the fourth ventricle into cerebellar corpus to form an anterior protrusion, which has been termed as cerebellar recessus (Kaslin et al., 2009). The latter grows further anteriorly thereby carrying the progenitors from the upper rhombic lip deeper into *corpus cerebelli*. Dye labelling experiments revealed that the lumen of cerebellar recessus extends across the cerebellar midline and remains in contact with the fourth ventricle even during adult stages. This shows that the nestin-expressing neuroepithelial-like precursors in the cerebellar stem cell niche stay in contact with the ventricular system with their apical sides, much like the precursors of subpallium and the periventricular tectal margin.

In our investigations, we verified the zones after injury with the neurogenic activity, the neurogenic niches located in the molecular layer of the dorsal, lateral, and basal areas of masu salmon's cerebellum. We linked their appearance to intensification of genetic proliferative programs in neutral stem cells and formation of local neurogenic niches as a response to damage. When damage is inflicted to neurogenic zones, active proliferation and differentiation of new cells begins, and they migrate to the damaged area, restoring the injured tissue.

According to Zupanc's data, in case of damage to *A. leptorhynchus* brain, intense neurogenesis processes start in the basal zone of cerebellum (Zupanc, 2011). Thus, this area can serve as a source of cells for reparative neurogenesis. In the basal zone of cerebellum in the area of granular eminences of juvenile masu salmon, on the contrary, the intensity of proliferation is high enough in the control and significantly decreases after a damaging effect. It is possible that the decrease of proliferation processes in this area of cerebellum is replaced by cell migration processes.

Appearance of neurogenic niches in the area of dorsal, lateral, and basal zones of the cerebellar molecular layer is the next stage of reparative neurogenesis. Neurogenic niches appear *de novo* after a traumatic impact and contain PCNA+ cells, as well as a heterogeneous population deprived of a proliferation marker. Immunolabeling of GFAP has shown that fibers of radial glia, as well as isolated small intensely and moderately labeled cells, which are phenotypically corresponding to neural stem cells described in cerebellum of other bony fish, are detected in the area of neurogenic niches (Hinsch and Zupanc, 2006; Kaslin et al., 2013; Sîrbulescu et al., 2015).

The investigations conducted allow us to conclude that, after inflicting an injury to cerebellum of juvenile masu salmon, processes of reparative neurogenesis are added to the processes of conctitutive neurogenesis: active proliferation and differentiation of new cells begin, and they migrate to the damaged area, repairing the tissue. The main source of new neurons is neurogenic niches that are formed during reparation of neural tissue.

GFAP, Doublecortin and other Markers of NSCs in Cerebellum of Fish

In cerebellum of juvenile masu salmon in normal state, we discovered a significant level of GFAP-positivity detected in fibers of systems of the spinocerebellar tract: as a part of cerebellar peduncles located in the basal part of cerebellar body, in fibers of radial glia, and in a heterogenous population of rounded (intensely labeled) and elongated (moderately labeled) cells (Stukaneva et al., 2017).

In the lateral zone, we discovered a small amount of DC+ cells of the following two types: small cells of 5.99 ± 0.11/2.76 ± 0.12 μm in size and larger ones of 9.07 ± 0.26/2.76 ± 0.34 μm in size. The number of intensely DC-labeled cells per visual section was no more than 15 (Figure 2A). They had a rounded or elongated shape, were deprived of process, and located solitary or formed small clusters (Figure 2B). In the granular layer of the basal zone, the probability density of differently directed, thin, DC+ radial glia fibers was moderate (Figure 2B). In a visual field, there were, on average, 14 DC+ cells and 24 DC+ radial glia cells. In the area of white matter of the basal zone (cerebellar peduncles), there were, on average, 30 DC+ cells and 48 GFAP+ fibers per a relevant test-field of 28 000 μm^2 (Stukaneva et al., 2017). In this area, fibers are prevailing GFAP+ elements, whose quantity differs significantly ($p < 0.05$). After injury, on the surface of the dorsal and basal zones and deep in the molecular layer of the lateral zone, we detected small local dense accumulations of GFAP- cells, whose morphological parameters correspond to those of cells of the first type labeled with PCNA (Stukaneva et al., 2017). We consider such GFAP+/DC+ cells conglomerates as neurogenic niches emerging in masu salmon's cerebellum as a response to damaging effect (Figure 2C, D). In dorsal, basal, and lateral zones, we found niches of 3192 ± 427, 6650 ± 721, and 1393 ± 653 μm^2, respectively (Stukaneva et al., 2017). In all the above-mentioned zones, parameters of proliferative PCNA+ cells coincide with the initial morphological parameters of GFAP+/DC+ cells.

Comparative studies of the ratio of GFAP+ cells to fibers in the dorsal and basal zones have been conducted. At one day after inflicting a damaging effect, we observed changes of cell composition and fibers compared to the intact state. After the damage, the number of GFAP+ cells and fibers increased in the dorsal and basal zones. We suppose that the increase in number of GFAP+ cells is connected with appearance of activated astrocytes as a result of the damage (Stukaneva et al., 2017).

As a result of the damage in the dorsal matrix area, we recorded appearance of GFAP+ fibers of radial glia forming differently oriented

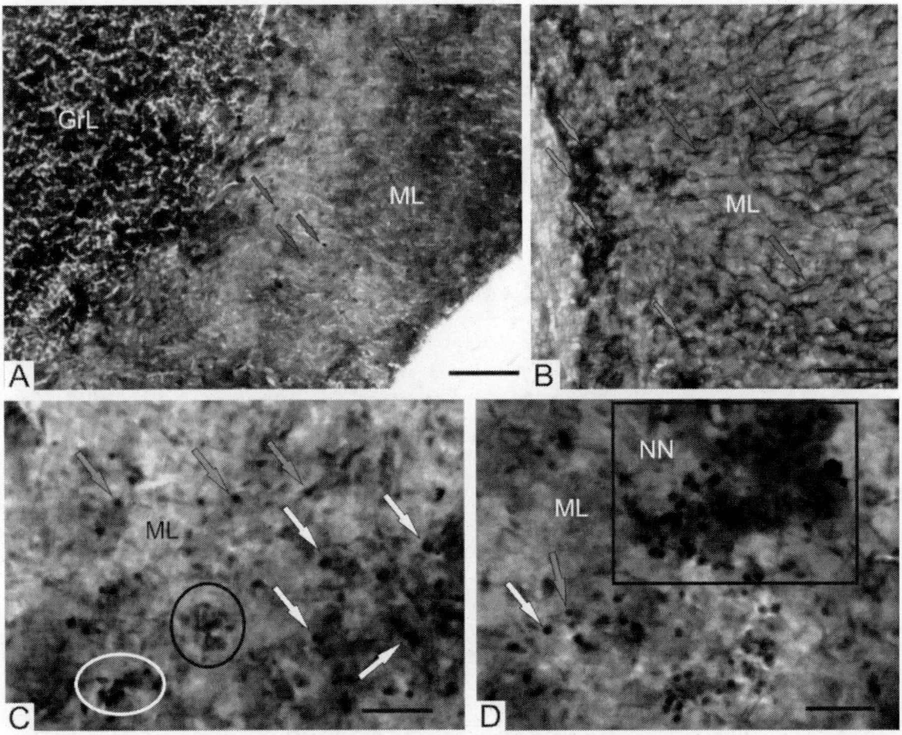

Figure 2. Expression of doublecortin (DC) in cerebellum of *Oncorhynchus masou* at 2 days post-injury: (A) general view of corpus cerebellum, (DC) immunopositive (ip) cells are indicated by pink arrows; (B) DC-ip cells (green arrows) and radial glia (blue arrows) in the basal part of corpus cerebellum; (C) DC-ip group of cells (in white oval), DC-ip single cells (white arrows) and DC- immunonegative (in) group of cells (in black oval) and DC-in single cells (orange arrows) in molecular layer of cerebellum; (D) a neurogenic niche containing DC-ip cluster of neural stem cells (in rectangle) in the molecular layer. Scale bar: A, 100 μm; B–D, 50 μm.

bundles (Figure 3). Such structural formations have not been detected in intact animals. We suppose that, after the injury in the dorsal matrix area, there is a structural reconstruction, connected with appearance of additional labeled fibers in the area of the damage, that leads to a twofold increase in the probability density of GFAP+ fibers in this area. We link such effects to intensification of spatial reorientation of radial glia fibers and formation of specific tracks for cells formed in this area.

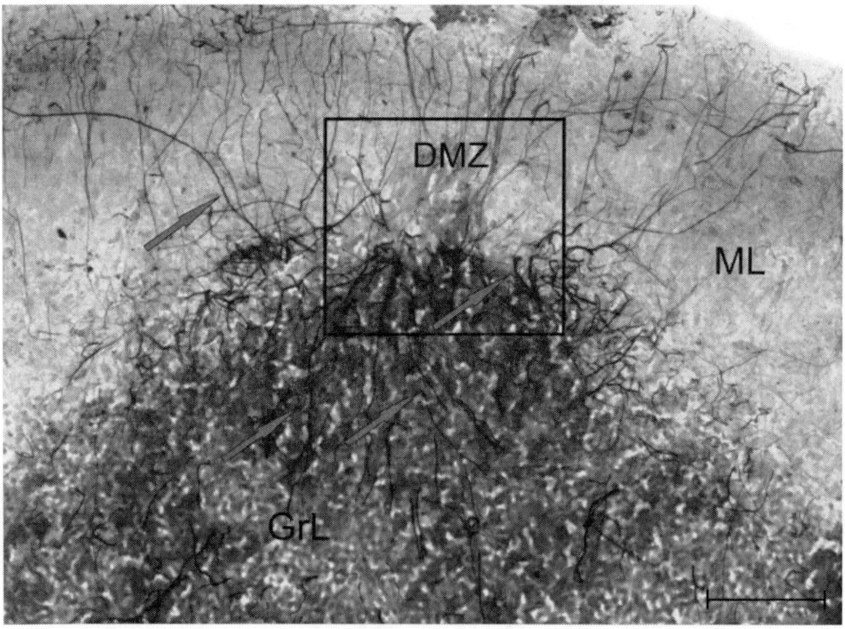

Figure 3. Localization of glial fibrillar acid protein (GFAP) in the cerebellum of *Oncorhynchus masou* at 2 days post-injury. Dorsal matrix zone is outlined by black squares; GFAP-immunopositive fiber is indicated by red arrows; DMZ, dorsal matrix zone; ML, molecular layer; GrL, granular layer. Scale bar: 100 μm.

In brain of adult mammals, cells of radial glia are located in restricted neurogenic areas, such as the subgranular zone of hippocampal dentate gyrus (Morrens et al., 2012; Dimou and Goetz, 2014). In mice, GFAP-labeled glia disappears from the beginning of myelinization (Nakahara et al., 2003), whereas in fish, GFAP-labeled fibers of radial glia are identified in different segments of the brain, including cerebellum, throughout their life (Zupanc and Horschke, 1995).

We suppose that the existence of GFAP+ and DC+ radial glia in an intact brain of juvenile masu salmon reflects processes of persistent neurogenesis. The morphology of GFAP+ structures, their topography, and interactions with other structural components allow us to assume that they obviously are elements of threadlike fibers of masu salmon's cerebellum. This is, in particular, indicated by the morphology of these structures and the presence of microcytocsculpture along fibers and their terminals.

In mammal brain, GFAP is a classical marker of astrocytes with numerous radially oriented processes and terminal pedicles that are often formed on the surface of vessels (Doetsch and Scharff, 2001). It has recently been shown that GFAP is one of the markers of neural stem cells (Kriegstein and Alvarez-Buylla, 2009; Ahmed et al., 2012). However, as a rule, their phenotype is more complex and, besides GFAP, contains vimentin, nestin, S-100, Sox2 (Adolf et al., 2006; Sîrbulescu et al., 2015). In studies of several areas of brain of adult bony fish, we identified precursor cells possessing characteristics of radial glia (Chapouton et al., 2007; Rothenaigner et al., 2011). In this case, they were capable of self-renewal and creation of different cell types, which demonstrate characteristics of genuine stem cells (Rothenaigner et al., 2011). It is possible that at least some of the radial glial cells in the brain of adult *A. leptorhynchus* fish act as these precursors. Since GFAP+ astrocytes can participate in recovery of cells lost in case of injury in the subventricular zone (Doetsch et al., 1999), it is obvious that these cells have the capability of self-repair. According to the data of Ahmed et al., (2012) cortex astrocytes in mouse brain stop fission by the tenth day of postnatal development, and GFAP expression after injury becomes more pronounced compared to proliferation processes.

GFAP+ cells of masu salmon's radial glia located in the surface layer of molecular layer have small-sized somas and long unbranched processes. Thus, GFAP+ cells and fibers of radial glia are both structural and morphological components in the composition of cerebellum of juvenile masu salmon. Our data support the results obtained for *Austrolebias affinis*, *A. charrua* and *A. reicherti* (Fernandez et al., 2011), and *A. leptorhynchus* (Clint and Zupanc, 2001) concerning the presence of GFAP+ radial glia in cerebellum of bony fish. We are inclined to attribute the few identified GFAP+ cells to neural stem cells. This assumption is based on a combination of morphological and topographic properties of neural stem cells, their connection with neurogenic zones, and relationships with fibers of radial glia.

In zones of secondary neurogenesis of masu salmon's cerebellum, we identified GFAP+ fibers and cells of radial glia. Data obtained from different species of bony fish is the evidence that radial glia is the prevalent type of

GFAP-immunolabeled elements remaining in brain of adult fish. In addition to these fibers, whose numbers increase after injury, cell migration also takes place (Clint and Zupanc, 2001).

Data of Ahmed et al., (2012) have shown that, after a mechanical damage to mouse brain, endogenic GFAP+ progenitor cells are activated. Approximately 50% of cells forming neurospheres in hippocampus and 75% in cortex originate from GFAP-expressing cells. According to data of Sîrbulescu et al., (2015) GFAP in cerebellum of *A. leptorhynchus* labels neural stem cells/precursor cells (Sîrbulescu et al., 2015). This agrees with a hypothesis that such cells are supposedly a source of neurospheres after the damage (Ahmed et al., 2012). However, not all neurospheres originate from GFAP+ cells: their smaller part is precursors of GFAP- cells.

CONCLUSION

Neurogenesis during adult stages is found in species of all vertebrate classes, as it serves different needs. Adult neurogenesis is considered in the context of protracted or constitutive growth or neuronal turnover. Adult neurogenesis is observed within the same class or order, even within one species, to serve these different needs. Are there fundamental differences in the generation of new neurons that show different phenotypes and serve different functions? From a morphological viewpoint, most stem cell populations retain contact to the ventricular system. Structurally, they appear as neuroepithelial cells, radial glial or astroglial cell types. The different shapes of these progenitors have been suggested to be a secondary consequence of the architecture of the developing parenchyma overlying the ventricular stem cell zone of the embryo (Kriegstein and Alvarez-Buylla, 2009). We suggest that an important step in characterizing the degree of difference between various stem cells pools is to identify the pathways that control their activity. This could be a key to our understanding of stem cell diversity and a valuable cue to the development of therapies based on yielding activation of active or dormant populations.

Certainly, in this scheme, SGZ radial astrocytes of mammalian brain are an odd element, as they are described as having no contact with the ventricle, yet they express nestin as do their cerebellar counterparts in zebrafish. This difference may be related to the late appearance of dentate gyrus in evolution which can explain the lack of a directly homologous cell type in other vertebrate groups (Kempermann, 2012). Progenitors of *eminentia granularis lateralis* have lost the ventricular contact also. It can be argued that they "only" serve the protracted cerebellar growth; however, SVZ astrocytes also serve the protracted growth of the olfactory bulb initially.

The differences between various stem cells on the level of gene expression may highlight their restriction and differences with respect to their prospective progeny. It will be interesting to find whether distinct stem cells can be forced to produce unusual progeny if needed. Such a scenario and, hence, a broader prospective potential of adult neural stem cells are confirmed by the lesion experiments in zebrafish cerebellum, which show, at least topologically, that newborn neurons will populate regions within the granular layer that are otherwise not attained. On the other hand, it still remains unclear if all the cell types that are lost after the injury are subsequently replaced indeed.

Funding

This work was supported by a grant from the President of the Russian Federation (No. MD-4318.2015.4), a grant from the Program for Basic Research of the Far East Branch of the Russian Academy of Sciences 2015–2017 (No. 15-I-6-116, section III).

References

Adolf B., Chapouton P., Lam C.S., Topp S., Tannhäuser B., Strähle U., Götz M. and Bally-Cuif L. Conserved and acquired features of adult neurogenesis in zebrafish telencephalon. *J. Dev. Biol.*, 2006, 295, 278-293.

Ahmed A., Shtaya A.B., Zaben M.J., Owens E.V. Kiecker C. and Gray W.P. Endogenous GFAP-positive neural stem/progenitor cells in the postnatal mouse cortex are activated following traumatic brain injury. *J. Neurotrauma*, 2012, 29, 828-842.

Berg D.A., Kirkham M., Beljajeva A., Knapp D., Habermann B., Ryge J., Tanaka E.M and Simon A. Efficient regeneration by activation of neurogenesis in homeostatically quiescent regions of the adult vertebrate brain. *Development*, 2010, 137, 4127-4134.

Candal E., Anadón R., Bourrat F., Rodríguez-Moldes I. Cell proliferation in the developing and adult hindbrain and midbrain of trout and medaka (teleosts): a segmental approach. *Brain Res. Dev. Brain Res.*, 2005, 160, 157-175.

Cerveny K.L., Varga M. and Wilson S.W. Continued growth and circuit building in the anamniote visual system. *Dev. Neurobiol.*, 2012, 72, 328-345.

Chaplin N., Tendeng C. and Wingate R.J. Absence of an external germinal layer in zebrafish and shark reveals a distinct, anamniote ground plan of cerebellum development. *J. Neurosci.*, 2010, 30, 3048-3057.

Chapouton P., Jagasia R. and Bally-Cuif L. Adult neurogenesis in non-mammalian vertebrates. *BioEssays*, 2007, 29, 745-757.

Clint S.C. and Zupanc G.K. Neuronal regeneration in the cerebellum of adult teleost fish, *Apteronotus leptorhynchus*: guidance of migrating young cells by radial glia. *Brain Res. Dev. Brain Res.*, 2001, 130, 15-23.

Curtis M.A., Kam M., Nannmark U., Anderson M.F., Axell M.Z., Wikkelso C., Holtas S., van Roon-Mom W.M., Björk-Eriksson T., Nordborg C., Frisén J. and Dragunow M. Human neuroblasts migrate to the olfactory bulb via a lateral ventricular extension. *Science*, 2007, 315, 1243-1249.

Dimou L. and Goetz M. Glial cells as progenitors and stem cells: new roles in the healthy and diseased brain. *Physiol. Rev.*, 2014, 94, 709-737.

Diotel N., Vaillant C., Gabbero C., Mironov S., Fostier A., Gueguen M.M., Anglade I., Kah O. and Pellegrini E. Effects of estradiol in adult neurogenesis and brain repair in zebrafish. *Horm. Behav.*, 2013, 63, 193-207.

Doetsch F., Caillé I. and Lim D.A. Subventricular zone astrocytes are neural stem cells in the adult mammalian brain. *Cell*, 1999, 97, 703-716.

Doetsch F. and Scharff C. Challenges for brain repair: insight from adult neurogenesis in birds and mammals. *Brain Behav. Evol.*, 2001, 58, 306-322.

Ekström P., Johnsson C.M. and Ohlin L.M. Ventricular proliferation zones in the brain of an adult teleost fish and their relation to neuromeres and migration (secondary matrix) zones. *J. Comp. Neurol.*, 2001, 436, 92-110.

Fernández A.S., Rocillo J.C., Casanova G. and Olivera-Bravo S. Proliferation zones in the brain of adult fish austrolebias (Cyprinodontiform: Rivulidae): a comparative study. *Neuroscience*, 2011, 189, 12-24.

Grandel H. and Brand M. Comparative aspects of adult neural stem cell activity in vertebrates. *Dev. Genes Evol.*, 2013, 223, 131-147.

Hinsch K. and Zupanc G.K. Isolation, cultivation, and differentiation of neural stem cells from adult fish brain. *J. Neurosci. Methods*, 2006, 158, 75-88.

Kaslin J., Ganz J. and Brand M. Proliferation, neurogenesis and regeneration in the nonmammalian vertebrate brain. *Philos. Trans. R. Soc. Lond. B Biol. Sci.*, 2008, 363, 101-122.

Kaslin J., Ganz J., Geffarth M., Grandel H., Hans S. and Brand M. Stem cells in the adult zebrafish cerebellum: initiation and maintenance of a novel stem cell niche. *J. Neurosci.*, 2009, 29, 6142-6153.

Kaslin J., Kroehne V., Benato F., Argenton F. and Brand M. Development and specification of cerebellar stem and progenitor cells in zebrafish: from embryo to adult. *Neural. Dev.*, 2013, 8:9, doi 10.1186/1749-8104-8-9.

Kempermann G. The neurogenic reserve hypothesis: what is adult hippocampal neurogenesis good for? *Trends Neurosci.*, 2008, 31, 163-169.

Kempermann G. New neurons for 'survival of the fittest'. *Nat. Rev. Neurosci.,* 2012, 13, 727-736.

Kirsche W. On postembryonic matrix zones in the brain of various vertebrates and their relationship to the study of the brain structure. *Z. Mikrosk. Anat. Forsch*., 1967, 77, 313-406.

Komada N. Distribution of taste buds in the oropharyngeal cavity of larval and juvenile stages of the cyprinid fish, *Tribolodon hakonensis*. *Jpn. J. Ichthyol*., 1994, 41, 307-311.

Kriegstein A. and Alvarez-Buylla A. The glial nature of embryonic and adult neural stem cells. *Annu. Rev. Neurosci*., 2009, 32, 149-184.

Kroehne V., Freudenreich D., Hans S., Kaslin J. and Brand M. Regeneration of the adult zebrafish brain from neurogenic radial glia-type progenitors. *Development*, 2011, 138, 4831-4841.

Kuroyanagi Y., Okuyama T., Suehiro Y., Imada H., Shimada A., Naruse K., Takeda H., Kubo T. and Takeuchi H. Proliferation zones in adult medaka (*Oryzias latipes*) brain. *Brain Res*., 2010, 1323, 33-40.

Levine E.M., Hitchcock P.F., Glasgow E. and Schechter N. Restricted expression of a new paired-class homeobox gene in normal and regenerating adult gold fish retina. *J. Comp. Neurol*., 1994, 348, 596-606.

Margotta V., Antonio M. and Enrico G. PCNA-positive cerebellar cells in an adult teleost, *Carassius carassius*. *J. Neurobiologia*, 2004, 15, 43-55.

März M., Schmidt R.S., Rastegar U. and Strähle U. Regenerative response following stab injury in the adult zebrafish telencephalon. *Dev. Dyn*., 2011, 240, 2221-2231.

Morales D. and Hatten M.E. Molecular markers of neuronal progenitors in the embryonic cerebellar anlage. *J. Neurosci*., 2006, 26, 12226-12236.

Morrens J., Van Den Broeck W. and Kempermann G. Glial cells in adult neurogenesis. *Glia*, 2012, 60, 159-174.

Nakahara J., Takemura M., Gomi H., Tsunematsu K., Itohara S., Asou H., Ogawa M., Aiso S. and Tan-Takeuchi K. Role of radial fibers in controlling the onset of myelination. *J. Neurosci. Res.,* 2003, 72, 279-289.

Noctor S.C., Flint A.C., Weissman T.A., Wong W.S., Clinton B.K. and Kriegstein A.R. Dividing precursor cells of the embryonic cortical ventricular zone have morphological and molecular characteristics of radial glia. *J. Neurosci.*, 2002, 22, 3161-3173.

Onteniente B., Kimura H. and Maeda T. Comparative study of the glial fibrillary acidic protein in vertebrates by pap immunohistochemistry. *J. Comp. Neurol.*, 1983, 215, 427-436.

Pellegrini E., Mouriec K., Anglade I., Menuet A., Le Page Y., Gueguen M.M., Marmignon M.H., Brion F., Pakdel F. and Kah O. Identification of aromatase-positive radial glial cells as progenitor cells in the ventricular layer of the forebrain in zebrafish. *J. Comp. Neurol.*, 2007, 501, 150-167.

Pouwels E. On the development of the cerebellum of the trout, *Salmo gairdneri*. I. Patterns of cell migration. *Anat. Embryol.*, 1978, 152, 291-308.

Raucci F., Di Fiore M.M., Pinelli C., D'Aniello B., Luongo L., Polese G. and Rastogi R.K. Proliferative activity in the frog brain: a PCNA-immunohistochemistry analysis. *J. Chem. Neuroanat.*, 2006, 32, 127-142.

Rothenaigner I., Krecsmarik M., Hayes J.A., Bahn B., Lepier A., Fortin G., Götz M., Jagasia R. and Bally-Cuif L. Clonal analysis by distinct viral vectors identifies bona fide neural stem cells in the adult zebrafish telencephalon and characterizes their division properties and fate. *Development*, 2011, 138, 1459-1469.

Seri B., García-Verdugo J.M., Collado-Morente L., McEwen B.S. and Alvarez-Buylla A. Cell types, lineage, and architecture of the germinal zone in the adult dentate gyrus. *J. Comp. Neurol.*, 2004, 478, 359-378.

Simmons A.M., Tanyu L.H., Horowitz S.S., Chapman J.A. and Brown R.A. Developmental and regional patterns of GAP-43 immuno-reactivity in a metamorphosing brain. *Brain Behav. Evol.*, 2008, 714, 247-262.

Sîrbulescu R.F., Ilieş I., Vitalo A.G., Trull K., Zhu J., Traniello I.M. and Zupanc G.K. Adult stem cells in the knifefish cerebellum. *Dev. Neurobiol.*, 2015, 75, 39-65.

Stukaneva M.E., Pushchina E.V., Varaksin A.A. and Shukla S. Monitoring of cell migration and apoptosis in cerebellum of juvenile masu salmon *Oncorhynchus masou* after injury. *Amer. J. BioSci.,* 2015, vol. 3, pp. 19-27.

Stukaneva M.E., Pushchina E.V. and Varaksin A.A. GFAP and PCNA marking in the cerebellum of masu salmon's (*Oncorhynchus masou*) juvenile after mechanical injury. *Rus. J. Dev. Biol.*, 2017, 48, 321-329.

Takeda A., Nakano M., Goris R.C. and Funakoshi K. Adult neurogenesis with 5-HT expression in lesioned goldfish spinal cord. *Neuroscience*, 2008, 151, 1132-1141.

Tozzini E.T., Baumgart M., Battistoni G. and Cellerino A. Adult neurogenesis in the short-lived teleost *Nothobranchius furzeri*: localization of neurogenic niches, molecular characterization and effects of aging. *Aging Cell*, 2011, 11, 241-251.

Wullimann M.F., Mueller T., Distel M., Babaryka A., Grothe B. and Köster R.W. The long adventurous journey of rhombic lip cells in jawed vertebrates: a comparative developmental analysis. *Front. Neuroanat.*, 2011, 5:27, doi: 10.3389/fnana.2011.00027.

Zupanc G.K. Neurogenesis and neuronal regeneration in the adult fish brain. *J. Comp. Physiol. A Neuroethol. Sens. Neural. Behav. Physiol.*, 2006, 192, 649-670.

Zupanc G.K. Adult neurogenesis in teleost fish. In: *Neurogenesis in the Adult Brain*. Eds. Seki T., Sawamoto K. and Parent J.M. Tokyo: Springer, 2011, vol. 1, pp. 137-168.

Zupanc G.K. and Horschke I. Proliferation zones in the brain of adult gymnotiform fish: a quantitative mapping study. *J. Comp. Neurol.*, 1995, 353, 213-233.

Zupanc G.K. and Ott R. Cell proliferation after lesions in the cerebellum of adult teleost fish: time course, origin, and type of new cells produced. *J. Exp. Neurol.*, 1999, 160, 78-87.

Zupanc G.K. and Sirbulescu R.F. Adult neurogenesis and neuronal regeneration in the central nervous system of teleost fish. *Eur. J. Neurosci.*, 2011, 34, 917-929.

Zupanc G.K. and Sirbulescu R.F. Teleost fish as a model system to study successful regeneration of the central nervous system. *Curr. Top. Microbiol. Immunol.*, 2013, 367, 193-233.

Zupanc G.K. and Zupanc M.M. New neurons for the injured brain: mechanisms of neuronal regeneration in adult teleost fish. *Regen. Med.*, 2006, 1, 207-216.

Zupanc G.K., Clint S.C., Takimoto N., Hughes A.T., Wellbrock U.M. and Meissner D. Spatio-temporal distribution of microglia/macrophages during regeneration in the cerebellum of adult teleost fish, *Apteronotus leptorhynchus*: a quantitative analysis. *Brain Behav. Evol.,* 2003, 62, 31-42.

In: Teleosts
Editor: Mikael Herleif
ISBN: 978-1-53613-660-9

Chapter 4

THE TELEOST RETINA AS A MODEL FOR REGENERATIVE MEDICINE

***Guadalupe Álvarez-Hernán*[1]*, Irene Sánchez-López*[1]*, Ruth Bejarano-Escobar*[1]*, Teresa Pavón-Muñoz*[1]*, Joaquín Rodríguez-León*[2]*, Gervasio Martín-Partido*[1] *and Javier Francisco-Morcillo*[1]**

[1]Área de Biología Celular, Departamento de Anatomía, Biología Celular y Zoología, Facultad de Ciencias, Universidad de Extremadura, Badajoz, Spain

[2]Área de Anatomía y Embriología Humana, Departamento de Anatomía, Biología Celular y Zoología, Facultad de Medicina, Universidad de Extremadura, Badajoz, Spain

ABSTRACT

The retina of teleosts has become a powerful model system in which to study different aspects of development and regeneration. While retinal development is completed during embryogenesis in most classes of vertebrates, it continues in teleosts throughout much of the animal´s life.

The mature retina grows from two specialized niches that sustain retinal stem cells (RSC): (i) the ciliary marginal zone (CMZ), a narrow ring of tissue located in the most-peripheral region of the retina that is composed of stem cells and progenitor cells that generate new neurons added to the periphery of the growing neural tissue, and (ii) the Müller glial cells, the radial glia of the neural retina, which generate neuronal progenitor cells that continue to proliferate and migrate to the photoreceptor layer, where they become precursor cells that are committed to differentiating into rod photoreceptors. Additionally, in the teleost retina, Müller glia have neurogenic capacity and mediates regeneration when stimulated by acute damage, by re-entering the cell cycle and re-establishing retinal cell types.

It has been suggested that the regenerative response of retinal tissue to injuries diminishes through the course of evolution from fish to mammals. Thus, the addition of new neurons from RSC is diminished in chickens compared to that observed in fish and amphibians. In contrast, the neuro-regenerative capacity of RSC in mammals is minimal or non-existent and therefore, degenerative retinal diseases in humans are irreversible. There are many genes, including structural proteins, signaling pathways and transcription factors that are expressed by normal RSC in teleosts that may be related to maintaining progenitor-like properties. Understanding the combinatorial expression of genes involved in the stimulation of the formation of neurogenic RSC in teleostean experimental models might lead to new genetic treatments for retinal degenerative diseases in humans.

Adaptations of the Fish Eye

The fish eye has morphological, structural, and physiological adaptations to the aquatic medium (Bejarano-Escobar et al., 2014) (Figure 1). Fish normally present lateral eyes (Figure 1A) and they usually have a flattened cornea, usually thinner than that of terrestrial vertebrates. Most of the fishes cannot control the amount of light entering the eye due to the absence of eyelids and a fixed pupil aperture. They also have a highly developed reflecting surface in the choroid that is called the *tapetum lucidum,* and a high refractive spherical lens (Figure 1B,C). The lens in teleosts is pulled back towards the retina by muscles called retractors, whereas in some chondrychthyes it is pulled away from the retina by muscles called protractors. Fish vision also shows adaptation to particular photoenvironments. Thus, deep sea fishes have an all-rod retina or a rod-dominated retina with a few cones sustaining a monochromatic vision. On the contrary, cones are generally found in diurnal active species. However, most of the fish species have color vision and generally have rod cells and cone cells.

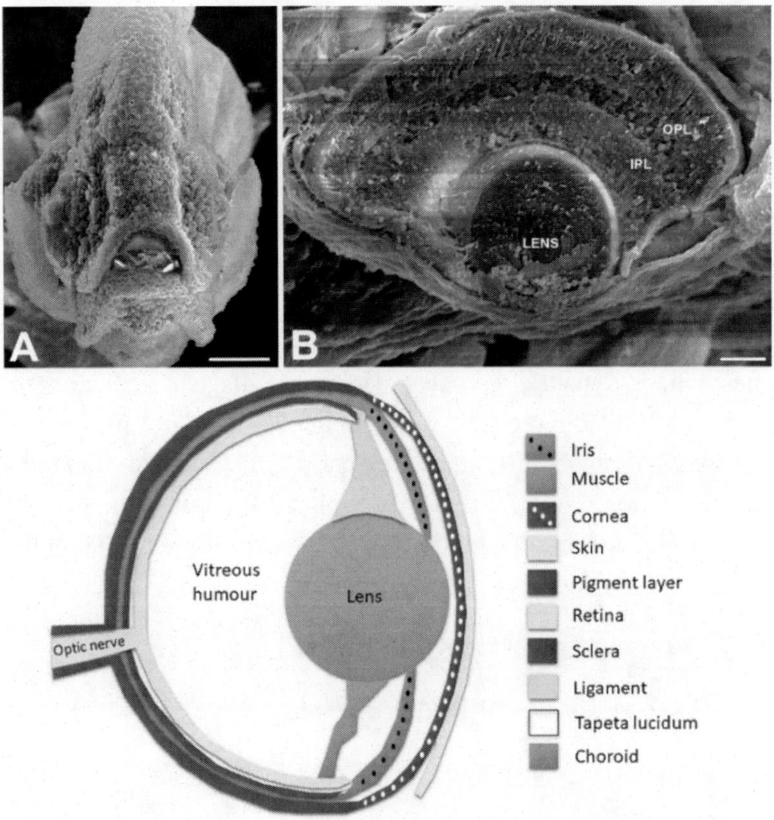

Figure 1. Structures and adaptations of the fish eye.A) Scanning electron microscopy of a *Solea senegalensis* symmetrical premetamorphic larvae. B) Section of the eye of *S. senegalensis* that shows the typical multi-layered structure of the retina. C) Scheme of the structures of the fish eye. One of the most notorious adaptations in the fish eye is the shape of the lens. As it can be observed in picture B, the lens has a spherical shape. *IPL, inner plexiform layer; OPL, outer plexiform layer.* Scale bars: 200 µm (A), 100 µm (B).

The FISH Retina

The fish retina is an excellent model in which to study developmental and physiological events. The neural retina is a thin layered structure whose main function consists on transducing light into nerve signals. It is located at the back of the eye upon which the visual image is focused by the cornea and lens. The retina is part of the central nervous system (CNS), originated as an outgrowth of the embryonic prosencephalon (see below). The neural retina is organized into different layers (Figure 2) which are, from the outermost to the innermost region: the outer nuclear layer (ONL), that is located next to the retinal pigment epithelium (RPE); the outer plexiform layer (OPL); the inner nuclear layer (INL); the inner plexiform layer (INL), the ganglion cell layer (GCL) and the optic fiber layer (OFL) (Figure 2). Nuclear layers are those containing cell bodies whereas plexiform layers are those ones where synaptic contacts between cells occur. The retina is constituted by six different types of neuronal cells that are: photoreceptors (cones and rods), bipolar cells, horizontal cells, amacrine cells and ganglion cells. Moreover, one type of radial glial cell can be found, the Müller glia (MG) (Figure 2B). Accessory glial cell types, such as microglial cells, can also be found in the teleost retina (Bejarano-Escobar et al., 2012b).

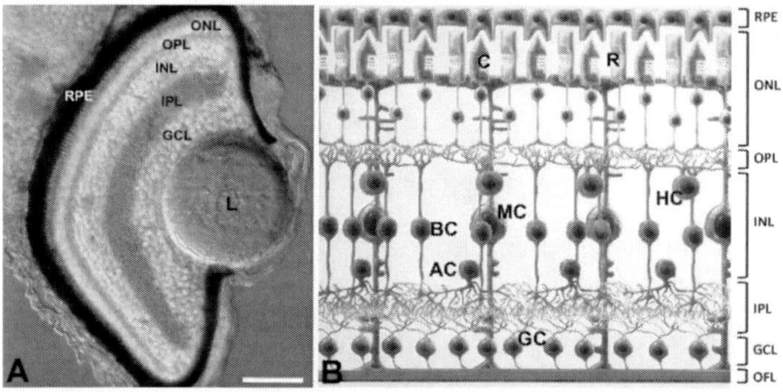

Figure 2. Histology and cell types of the fish retina. A) Retinal section that shows the multi-layered structure of mature retina. Notice the spherical lens. *GCL, ganglion cell layer; INL, inner nuclear layer; IPL, inner plexiform layer; L, lens; ONL, outer nuclear layer; OPL, outer plexiform layer; PE, pigmented epithelium.* B) Schematic representation of retinal cell types. *AC, amacrine cell; BC, bipolar cell; C, cone; GC, ganglion cell; HC, horizontal cell; MC, Müller cell; R, rod.*Scale bar: 100 μm.

Fish Eye Development and Retinogenesis

Eye development follows a common pattern in all vertebrate animals (Figure 3). Eye morphogenesis occurs early in the developing fish embryo. It starts with a bilateral evagination of the diencephalon on the early neurula that leads to the formation of the optic vesicles (Figure 3A). These extend towards the surface ectoderm through the adjacent mesenchyme. Then, the vesicles come into contact with the ectoderm, displacing the mesenchyme. Concomitantly, the surface ectoderm thickens forming the lens placode (Figure 3B). Both lens placode and optic vesicle invaginate, becoming the lens vesicle and a doubled-layered structure called the optic cup, respectively (Figure 3C). The inner layer of the optic cup forms the neural retina, while the outer layer becomes the RPE (Figure 3D).

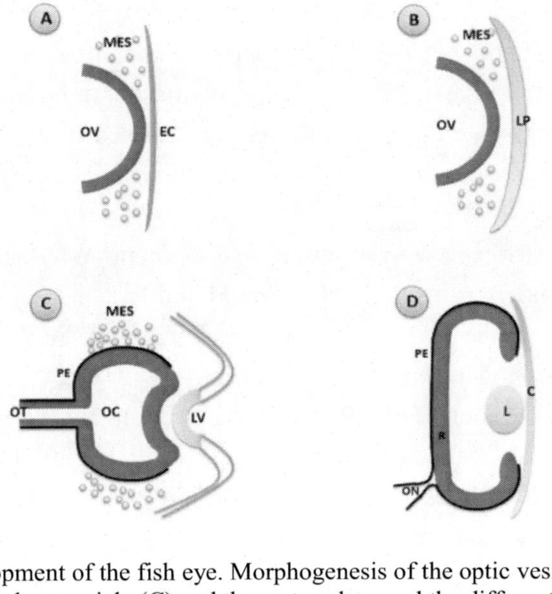

Figure 3. Development of the fish eye. Morphogenesis of the optic vesicle (A, B), the optic cup (C), the lens vesicle (C) and the mature lens and the differentiated retina (D). *C, cornea; Ec, ectoderm; L, lens; LP, lens placode; LV, lens vesicle; Mes, mesenchyme; OC, optic cup; ON, optic nerve; OT, optic tract; OV, optic vesicle; PE, pigmented epithelium; R, retina.*

Concerning the process of cell differentiation, RPE cells synthetize melanin. On the other hand, proliferating neuroepithelial cells are found in the presumptive neural retina (Figure 4A,E). The nuclei of these retinal stem cells migrate in an apical-to-basal manner and in phase with the cell cycle (Figure 4E). Thus, abundant apical mitoses are always found at the ventricular surface of the retina by these initial stages of eye development. As development proceeds, asymmetrical cell divisions produce migratory neuroblasts that lose contact with the ventricular surface of the retina and migrate radially to their final location, in which they differentiate (Figure 4D). During stages previous to the beginning of cell differentiation, retinal rudiment is composed of a neuroblastic layer (NbL) (Figure 4B). As cell differentiation occurs, the emergence of the retinal layers can also be observed (Figure 4C), increasing in thickness with the progression of retinal maturity (Figure 4D).

This change in the potential of retinal cells is likely to be due to a combination of intrinsic and extrinsic factors, such as transcription factors and signaling molecules, respectively (Xiang, 2013). The onset of cell differentiation can be determined by using retinal markers to show the cessation of proliferative activity in the retinal tissue, or by employing cell-type-specific markers (for a review, see Bejarano-Escobar et al., 2014). This process will result in the formation of all neural retinal cell types described above (Figure 4E).

Fish species differ in their mode of development. Whereas in precocial species the specimens skip the larval period and exit the egg in a juvenile state, altricial prolarvae exit the egg in an undeveloped state. Concerning to visual system, retinogenesis is completed during the embryological period in precocial species, such as goldfish (Sharma and Ungar, 1980), brown trout (Candal et al., 2005), medaka (Kitambi and Malicki, 2008), and several species of chondrichthyes (Harahush et al., 2009; Ferreiro-Galve et al., 2010; Bejarano-Escobar et al., 2012a; 2013) (Figure 5A,B). However, altricial species are born blind, and the acquisition of the full adult complement of cells does not occur until several days, weeks or even months after hatching, as occurs in the tench (Bejarano-Escobar et al., 2009), Senegalese sole (Bejarano-Escobar et al., 2010), or the gilthead seabream (Pavón-Muñoz et al., 2016) (Figure 5C,D).

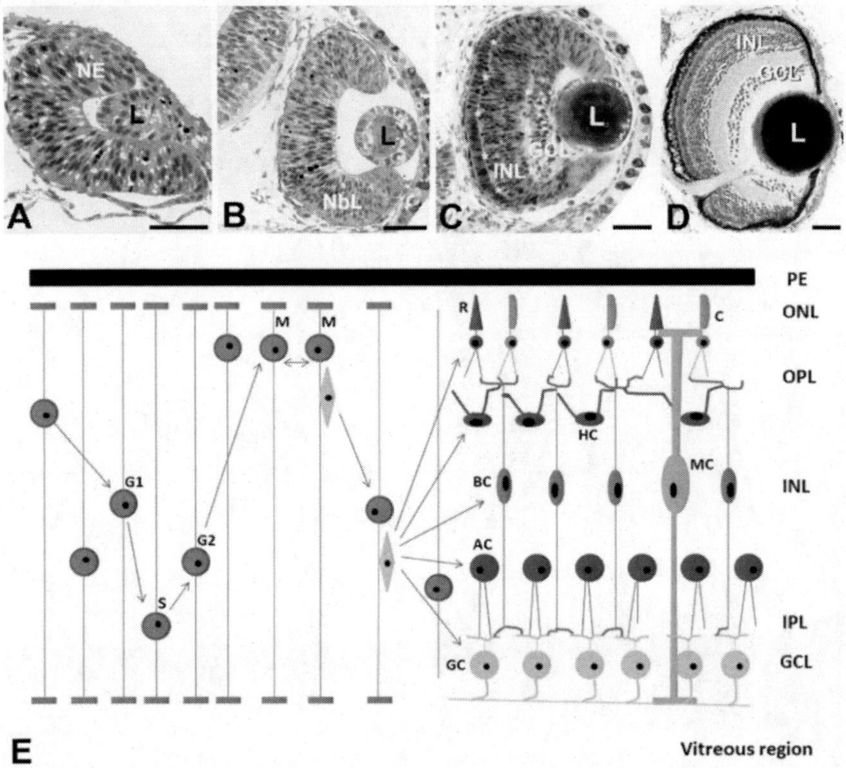

Figure 4. Fish retinogenesis. A-D) Toluidine-blue stained sections show the different phases of development of the retina. A) The retina is composed of a pseudostratified epithelium of multipotent progenitor cells. B) The retina is constituted of a neuroblastic layer. At this stage, neuroblastic cells migrate along the tissue to be located on its final position. C) Emergence of the nuclear and plexiform layers (asterisks). D) Mature retina shows a typical multi-layered structure. E) Cell differentiation. Migratory neuroblasts move from the vitreous (S phase) to the ventricular (M phase) region. When these cell types leave the cell cycle, the process of differentiation begins. *AC, amacrine cells; BC, bipolar cells; C, cones; GC, ganglion cells; GCL, ganglion cell layer; HC, horizontal cells; INL, inner nuclear layer; IPL, inner plexiform layer; L, lens; MC, Müller cells; NbL, neuroblastic layer; NE, neuroepithelium; ONL, outer nuclear layer; OPL, outer plexiform layer; PE, pigmented epithelium; R, rodes.* Scale bars: 100 μm.

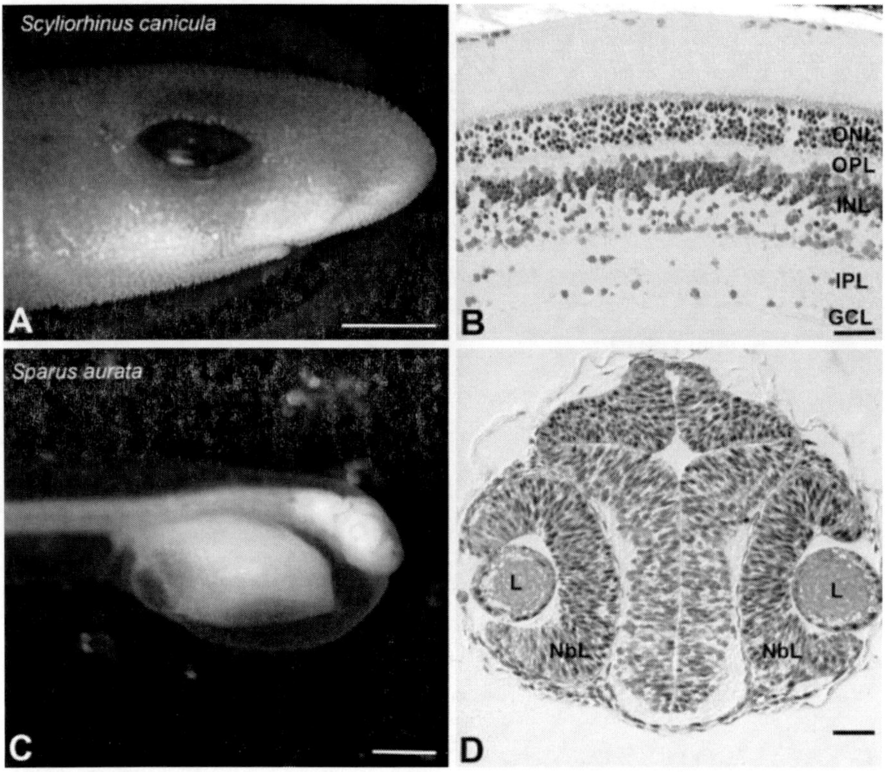

Figure 5. Retinal maturity at hatching in precocial (A, B) and altricial (C,D) fish species. At the hatching day, the retina of altricial species is still undifferentiated and is composed of a neuroblastic layer (A, B) while the retina of precocial species shows a completely differentiated tissue (C, D). Scale bars: 10 mm (A), 500 µm (C), 100 µm (B, D).

In any case, the retina of both precocial and altricial teleost species presents a very intense postnatal neurogenesis. Thus, as the fish grows, the retina continues to grow throughout its life and new neurons are constantly added to the mature retinal tissue. It is now widely recognized that retinal stem cells persist in specialized "niches" in the adult teleost retina where they generate a large number of different types of neurons (Raymond et al., 2006, Stenkamp, 2011; Lenkowski and Raymond, 2014).

Stem Cell Niches in the Teleost Retina

Healthy Retina

The proliferating cells in the mature fish retina are shown to accumulate ^{3}H thymidine (Raymond and Rivlin, 1987) or BrdU (Julian et al., 1998; Otteson et al., 2001). They can also be identified immunohistochemically with antibodies against the proliferating cell nuclear antigen (PCNA) (Bejarano-Escobar et al., 2009; 2010; 2014; Pavón-Muñoz et al., 2016) (Figure 6B,D,F,H) and against phospho-histone H3 (pHisH3) (Pavón-Muñoz et al., 2016) (Figure 6C,D,G,H). Proliferative activity is mainly detected in the ciliary marginal zone (CMZ), a circumferential ring of cells located at the periphery of the neural retina (Figure 6). Furthermore, proliferative Müller cells are also detected in the INL. These cells continue to divide and likely give rise to the precursor cells that migrate to the ONL and, after several cycles of cell divisions, differentiate into new rod photoreceptors (Figure 6). These stem cell niches are detailed below.

Ciliary Marginal Zone (CMZ)

The CMZ is also known in other vertebrates as *ora serrata*, *ora terminalis*, or circumferential germinative zone (CGZ) (Amato et al., 2004; Fischer et al., 2014). The CMZ is a remnant of the embryonic neuroepithelium preserved at the retinal margin following cell differentiation and lamination centrally (Raymond et al., 2006). It has been shown that new rings of cells are constantly added from the teleost CMZ to the mature tissue. Therefore, cells situated more peripherally are younger than central ones. The teleost CMZ is able to generate an enormous number of progeny, since the total number of retinal cells generated far exceeds the number of CMZ cells present at the end of embryogenesis (Figure 7). However, reptilian and avian retinal histogenesis takes place almost exclusively during embryonic stages (Figure 7). Proliferative activity is detected in the CMZ of younger chicks (<30 days) and few retinal tissue is generated in the peripheral region in hatchlings (Fischer and Reh, 2000). The proliferation of these undifferentiated cells located in the CMZ can be greatly enhanced post-natally by intraocular injections of EGF, IGF-I, or insulin (Fischer and Reh, 2000). On the contrary, mammals do not show evidence of proliferative activity at the peripheral edge of the normal retina at birth, suggesting that mammals do not have CMZ (Figure 7).

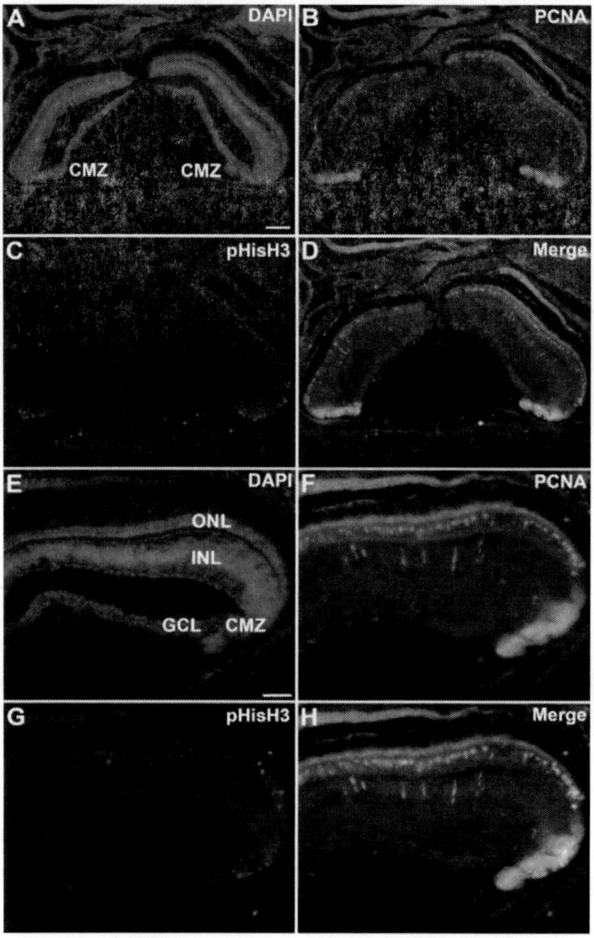

Figure 6. Niches of stem cells in mature retina.pHisH3 and PCNA-immunoreactivities of *Sparus aurata* retina. Magnifications of A-D are shown in E-H. Sections were counterstained with DAPI. The presence of the CMZ can be observed at the periphery of the retina (asterisks in B,D,F,H). At this region, CMZ is composed of retinal stem cells and retinal progenitor cells that generate retinal neurons. Anti-PCNA antibody shows the presence proliferative photoreceptors (arrows in B,D,F,H) that originate from proliferative Müller cells (B,D,F,H). Anti-pHisH3 antibody shows the presence of mitotic activity in the CMZ and in the photoreceptor layer (arrowheads in C,D,G,H). Scale bars: 200 µm in A (A-D), 100 µm in E (E-H).

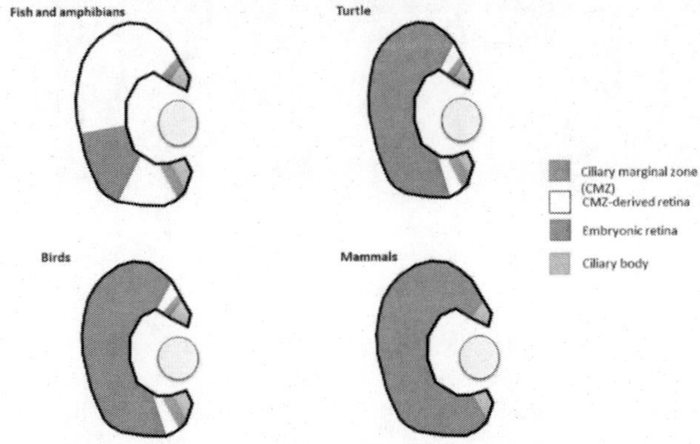

Figure 7. Schematic representation of the presence and absence of the CMZ among the vertebrates and its contribution to retinal growth. Neurogenesis continues after embryonic stages. In fish and amphibians, most of the retina derives from the CMZ, which contributes to the regeneration of this tissue. The presence of the CMZ is conserved in turtles and birds but its contribution to the generation of the retina is minimal. Most of this tissue is embryonic. In mammals, the CMZ is absent and the whole retina originates from embryonic ectoderm. (*Adapted from Moshiri et al., 2004)*.

In order to determine the histological organization of the CMZ, Raymond et al., (2006) systematically examined the expression patterns of selected genes involved in retinogenesis by using *in situ* hybridization. They subdivided the zebrafish CMZ into four anatomical regions: peripheral, middle, central-outer, and central-inner. Retinal stem cells are located in the peripheral sub-region and more restricted retinal progenitors lay closer to the differentiated retina. Thus, the definitive GCL derives from the middle sub-region of the CMZ. Progenitors at the central-inner CMZ give rise to cells located in the INL, and the central-outer sub-region generates cone photoreceptors in the ONL (Raymond et al., 2006). More recently, the clonal analysis and imaging studies showed that, as the retina develops, retinal stem cells located in the extreme peripheral edge of the CMZ divide asymmetrically to generate a new retinal stem cell that remains in the stem cell niche, and a second daughter cell that is pushed centrally, where it becomes a more restricted retinal progenitor cell (Wan et al., 2016). Therefore, the CMZ is able to generate all the retinal cell types with the exception of rods (Figure 8) that are generated by the Müller glia (see below).

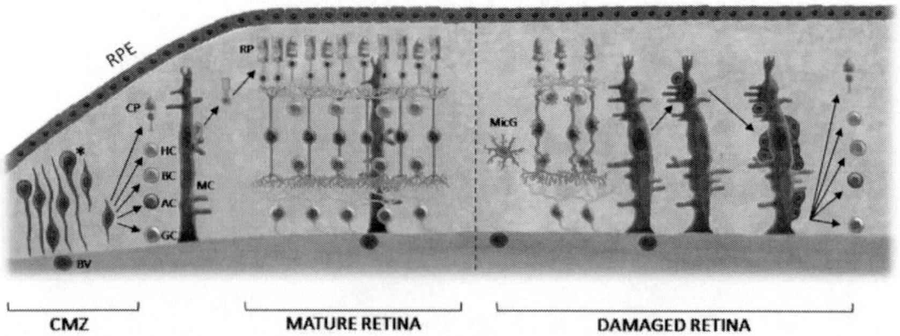

Figure 8. Neurogenesis of the fish retina during growth and regeneration. The CMZ is composed of retinal stem cells and retinal progenitor cells. RPCs are located in the peripheral retina and they will differentiate into retinal neurons. Müller cells are retinal progenitor cells that give rise to rodes. When the retina is damaged, microglial cells are activated and remove cellular debris. Nuclei of Müller cells move to the apical surface and divide asymmetrically. This process results in the formation of RPCs that migrate to the right position to generate the typical retinal neurons. (*Adapted from Lenkowski and Raymond, 2014*).

Müller Glia

In the normal developing and mature healthy retina, Müller cells have been characterized as the "radial glia" of the retina. Based on functional similarities, many of those authors considered MG as astrocyte-like cells. These cells are clearly defined by structure, function, and gene expression patterns (Reichenbach and Bringmann, 2013; Gallina et al., 2014; Lenkowski and Raymond, 2014). MG are involved in homeostasis, retinal innate immunity, retinal diseases and regeneration of the visual system, and phagocytosis of cell debris and diverse foreign bodies under physiological, pathological, and experimental conditions (Reichenbach and Bringmann, 2013; Gallina et al., 2014; Goldman, 2014; Gorsuch and Hyde, 2014; Lenkowski and Raymond, 2014; Hamon et al., 2016; Vecino et al., 2016; Bejarano-Escobar et al., 2017). However, within the past years, MG have been identified as a source of retinal cells in the healthy (Bernardos et al., 2007) and injured teleost retina (Fausset and Goldman, 2006). Thus, it has been shown that MG of teleosts continue to divide slowly and give rise to rod progenitors which form clusters in the INL that align in radial columns. These rod progenitors migrate away from the INL and into the

ONL, where they continue to divide and differentiate into new rod photoreceptors (Julian et al., 1998; Bernardos et al., 2007) (Figure 8).

Injured Retina: Regeneration

It has been shown that adult teleosts can regenerate their retinas after different methods of injury such as cytotoxic lesion by injection of neurotoxins (Fimbel et al., 2007; Sherpa et al., 2008), surgical approach (Hitchcock et al., 1992; Fausset and Goldman, 2006), localized heat (Raymond et al., 2006), laser or light damage (Vihtelic and Hyde, 2000; Bernardos et al., 2007; Thummel et al., 2008; Bejarano-Escobar et al., 2012b). In the case of constant intense-light exposure, cone-diurnal retina of albino fish (Allen et al., 1999; Vihtelic and Hyde, 2000; Allison et al., 2006), or rod-dominated retina of larval epibenthonic teleosts (Bejarano-Escobar et al., 2012b), are suitable models because of their relatively high susceptibility to light damage. In any case, after cell degeneration in the teleost retina, regeneration is conducted by the stem cells present at the niches described above. Thus, it has been demonstrated that retinal damage in teleosts causes increased cell proliferation at the peripheral edge, suggesting that the CMZ contributes to retinal regeneration (Raymond et al., 1988; Hitchcock et al., 1992). There is also evidence that after neurotoxic treatment, cones regenerate with a different pattern and density at the periphery than in the central goldfish retina (Stenkamp et al., 2001). This suggests that two cone populations arise via two spatially and cellularly distinct mechanisms. These authors suggest that the CMZ is the major source of the regenerated peripheral retina, but not of the central one. It was recently demonstrated that after rod cell ablation, CMZ stem cells generate all retinal cell types with the exception of rod photoreceptors (Wilson et al., 2016). It remains unclear why the proliferative response of the CMZ increases in rod degenerative fish models since they do not contribute to rod neurogenesis.

Although under normal physiological conditions, MG generate only rod photoreceptors, it is well established that a subset of Müller cells are the source of central retinal regeneration in teleosts (Fausset and Goldman, 2006; Bernardos et al., 2007; Thummel et al., 2008). Thus, after retinal damage, MG reenter the cell cycle and act similar to retinal progenitor cells during embryonic development, exhibiting inter kinetic nuclear migration apically to undergo a single asymmetric mitosis to self-renew and to produce a proliferating retinal progenitor (Nagashima et al., 2013). Nuclear migration is facilitated by the actin cytoskeleton and Rho-associated coiled-coil kinases (Rocks) and it is necessary for photoreceptor regeneration in the adult zebrafish retina (Lahne et al., 2015). The newly generated cell divides rapidly and repeatedly and a cluster of progenitor cells is raised. The vitreal or scleral migration of dividing progenitors and post-mitotic neural precursors is a necessary step to replace the population of cells that has been lost and restore the functional vision (Figure 8). Thus, the selective ablation of photoreceptors recruits progenitors to the ONL (Vihtelic and Hyde, 2000). In contrast, the ablation of cells located in the inner half of the retina retains the retinal progenitors in the INL to replace amacrine or ganglion cells (Fimbel et al., 2007) (Figure 8).

Different cytokines secreted after retinal damage in teleosts promote MG de-differentiation and proliferation by the expression of several proneural transcription factors. Thus, the pro-inflammatory cytokine TNFα produced by apoptotic neurons initiates MG proliferation (Nelson et al., 2013). Stat3, as well as members of the MAPK and Akt signaling pathways, increase their expression in MG prior to their de-differentiation and proliferative response (Kassen et al., 2007). Stat3-dependant signaling pathway is stimulated by ciliary neurotrophic factor (CNTF) (Kassen et al., 2009). The expression of the proneural basic-helix-loop-helix transcription factor Achaete-scute homolog 1a (Ascl1a) and the subsequent induction of lin-28, a pluripotency mRNA binding protein, is necessary for teleost MG and retinal regeneration (Ramachandran et al., 2010). These authors also showed that Müller glial cells respond by re-expressing pluripotency genes such as *c-myc, nanog, nox2, and oct4*. Other studies have shown that Ascl1a-Dkk-Wnt signaling pathway, in conjunction with Pax6, Stat3, and Hspd1, regulates MG de-differentiation and progenitor proliferation (Kassen et al., 2009; Qin et al., 2009; Thummel et al., 2010; Ramachandran et al., 2011). A recent transcriptomic analysis has revealed that several other categories of genes/signaling, such as nuclear factor-κB (NF-κB), clock genes, and prostaglandin metabolism, rapidly up-regulate in reactive MG (Sifuentes et al., 2016). More recently, it has been shown that sex determining region Y-box 2 (Sox2) (Gorsuch et al., 2017), and atonal homolog 7 (Atoh7) (Lust et al., 2016) are re-expressed in Müller cells after injury and lead them to proliferate and generate new retinal neurons. Additionally, Sox2 has the capacity to induce Ascl1a and Atoh7 expression, which are able to activate lin-28 expression themselves. It has also been shown that Notch signaling-inhibition collaborates with Ascl1a and Lin28a to stimulate widespread MG proliferation in the uninjured retina (Elsaeidi et al., 2018). Recently, microRNAs (miR) have also been identified to control the generation of new neurons derived from teleosts MG (Madelaine and Mourrain, 2017). Thus, the depletion of miR-9 increases the number of undifferentiated progenitors in the teleost retina. One of the transcription factors mentioned above, *lin-28,* which is involved in inducing proliferation of neural retinal stem cells after damage, is a target of miR-9 (La Torre et al., 2013). Other authors have shown that microglial cells and macrophages are critical for proper rod cell regeneration, stimulating the activation of retinal MG to a stem cell-like state (White et al., 2017).

Therapeutic Applications

As we have previously shown, the retina of the teleost fish has a remarkable potential to regenerate new neurons after various types of injury. Our understanding of the molecular mechanisms that regulate retinal regeneration from MG in fish has advanced rapidly during the last decade. Currently, the MG is recognized as the primary source of new neurons after retinal damage also in birds (Gallina et al., 2014).

Mammals do not generally show evidence of proliferation at the peripheral edge of the neural retina, suggesting that this group of vertebrates do not have CMZ. Therefore, the attempts to stimulate retinal regeneration in the mammalian retina have focused on the MG. Retinal injury results in reactive gliosis and glial scaring in mammals (Bringmann et al., 2009) and traditional studies have failed to identify any regenerative potential in the mammalian retina. Thus, degenerative retinal diseases in mammals can lead to vision loss. However, retinal injury in either fish or mammals induces the up-regulation of many mitogenic factors (Kassen et al., 2009; Wan et al., 2012; 2014). Therefore, the initial steps in the response to injury that lead to regeneration in teleosts are also conserved in the retina of mammals. Thus, *in vitro* studies have demonstrated that mammalian Müller cells could act as progenitors of bipolar (Pollak et al., 2013) and photoreceptor cells (Giannelli et al., 2011). More recently, it has been shown that forced expression of Ascl1 promotes MG proliferation and differentiation into amacrine, bipolar, and photoreceptor neurons following retinal injury in young mice (Ueki et al., 2015). The combination of Ascl1 expression with a histone deacetylase inhibitor increases the differentiation of MG into new neurons in the mouse adult retina (Jorstad et al., 2017).

Therefore, our understanding of the molecular mechanisms that regulate retinal regeneration from stem cells in teleosts could bring insights into the molecular mechanisms that either sustain or prevent neural cell replacement and therefore help to design therapies to stimulate retinal regeneration in mammals.

REFERENCES

Allen, D. M., Pipes, C., Deramus, K. & Hallows, T. E. (1999). *A Comparison of Light-Induced Rod Degeneration in Two Teleost Models Retinal Degenerative Diseases and Experimental Therapy.* (Hollyfield, J. G., Anderson, R. E. & LaVail, M. M., eds.), pp. 337-350: Springer US.

Allison, W. T., Hallows, T. E., Johnson, T., Hawryshyn, C. W. & Allen, D. M. (2006). Photic history modifies susceptibility to retinal damage in albino trout. *Vis Neurosci 23*, 25-34.

Amato, M. A., Arnault, E. & Perron, M. (2004). Retinal stem cells in vertebrates: parallels and divergences. *Int J Dev Biol 48*, 993-1001.

Bejarano-Escobar, R., Blasco, M., DeGrip, W. J., Martín-Partido, G. & Francisco-Morcillo, J. (2009). Cell differentiation in the retina of an epibenthonic teleost, the Tench (*Tinca tinca*, Linneo 1758). *Exp Eye Res 89*, 398-415.

Bejarano-Escobar, R., Blasco, M., DeGrip, W. J., Oyola-Velasco, J. A., Martín-Partido, G. & Francisco-Morcillo, J. (2010). Eye development and retinal differentiation in an altricial fish species, the senegalese sole (*Solea senegalensis*, Kaup 1858). *J Exp Zool B Mol Dev Evol 314*, 580-605.

Bejarano-Escobar, R., Blasco, M., Durán, A. C., Martín-Partido, G. & Francisco-Morcillo, J. (2013). Chronotopographical distribution patterns of cell death and of lectin-positive macrophages/microglial cells during the visual system ontogeny of the small-spotted catshark *Scyliorhinus canicula*. *J Anat 223*, 171-184.

Bejarano-Escobar, R., Blasco, M., Durán, A. C., Rodríguez, C., Martín-Partido, G. & Francisco-Morcillo, J. (2012a). Retinal histogenesis and cell differentiation in an elasmobranch species, the small-spotted catshark *Scyliorhinus canicula*. *J Anat 220*, 318-335.

Bejarano-Escobar, R., Blasco, M., Martín-Partido, G. & Francisco-Morcillo, J. (2012b). Light-induced degeneration and microglial response in the retina of an epibenthonic pigmented teleost: age-dependent photoreceptor susceptibility to cell death. *J Exp Biol. 215*. 3799-3812.

Bejarano-Escobar, R., Blasco, M., Martín-Partido, G. & Francisco-Morcillo, J. (2014). Molecular characterization of cell types in the developing, mature, and regenerating fish retina. *Reviews in Fish Biology and Fisheries 24*, 127-158.

Bejarano-Escobar, R., Sánchez-Calderón, H., Otero-Arenas, J., Martín-Partido, G. & Francisco-Morcillo, J. (2017). Müller glia and phagocytosis of cell debris in retinal tissue. *J Anat. 231,* 471-483.

Bernardos, R. L., Barthel, L. K., Meyers, J. R. & Raymond, P. A. (2007). Late-stage neuronal progenitors in the retina are radial Müller glia that function as retinal stem cells. *J Neurosci 27*, 7028-7040.

Bringmann, A., Iandiev, I., Pannicke, T., Wurm, A., Hollborn, M., Wiedemann, P., Osborne, N. N. & Reichenbach, A. (2009). Cellular signaling and factors involved in Muller cell gliosis: neuroprotective and detrimental effects. *Prog Retin Eye Res 28*, 423-451.

Candal, E., Anadón, R., DeGrip, W. J. & Rodríguez-Moldes, I. (2005). Patterns of cell proliferation and cell death in the developing retina and optic tectum of the brown trout. *Brain Res Dev Brain Res 154*, 101-119.

Elsaeidi, F., Macpherson, P., Mills, E. A., Jui, J., Flannery, J. G. & Goldman, D. (2018). Notch Suppression Collaborates with Ascl1 and Lin28 to Unleash a Regenerative Response in Fish Retina, But Not in Mice. *J Neurosci 38*, 2246-2261.

Fausett, B. V. & Goldman, D. (2006). A role for alpha1 tubulin-expressing Muller glia in regeneration of the injured zebrafish retina. *J Neurosci 26*, 6303-6313.

Ferreiro-Galve, S., Rodríguez-Moldes, I., Anadón, R. & Candal, E. (2010). Patterns of cell proliferation and rod photoreceptor differentiation in shark retinas. *J Chem Neuroanat 39*, 1-14.

Fimbel, S. M., Montgomery, J. E., Burket, C. T. & Hyde, D. R. (2007). Regeneration of inner retinal neurons after intravitreal injection of ouabain in zebrafish. *J Neurosci 27*, 1712-1724.

Fischer, A. J. & Reh, T. A. (2000). Identification of a proliferating marginal zone of retinal progenitors in postnatal chickens. *Dev Biol 220*, 197-210.

Fischer, A. J., Zelinka, C., Gallina, D., Scott, M. A. & Todd, L. (2014). Reactive microglia and macrophage facilitate the formation of Muller glia-derived retinal progenitors. *Glia 62*, 1608-1628.

Gallina, D., Todd, L. & Fischer, A. J. (2014). A comparative analysis of Müller glia-mediated regeneration in the vertebrate retina. *Exp Eye Res 123*, 121-130.

Giannelli, S. G., Demontis, G. C., Pertile, G., Rama, P. & Broccoli, V. (2011). Adult human Müller glia cells are a highly efficient source of rod photoreceptors. *Stem Cells 29*, 344-356.

Goldman, D. (2014). Müller glial cell reprogramming and retina regeneration. *Nat Rev Neurosci 15*, 431-442.

Gorsuch, R. A. & Hyde, D. R. (2014). Regulation of Müller glial dependent neuronal regeneration in the damaged adult zebrafish retina. *Exp Eye Res 123*, 131-140.

Gorsuch, R. A., Lahne, M., Yarka, C. E., Petravick, M. E., Li, J. & Hyde, D. R. (2017). Sox2 regulates Müller glia reprogramming and proliferation in the regenerating zebrafish retina via Lin28 and Ascl1a. *Exp Eye Res 161*, 174-192.

Hamon, A., Roger, J. E., Yang, X. J. & Perron, M. (2016). Müller glial cell-dependent regeneration of the neural retina: An overview across vertebrate model systems. *Dev Dyn 245*, 727-738.

Harahush, B. K., Hart, N. S., Green, K. & Collin, S. P. (2009). Retinal neurogenesis and ontogenetic changes in the visual system of the brown banded bamboo shark, *Chiloscyllium punctatum* (Hemiscyllidae, Elasmobranchii). *J Comp Neurol 513*, 83-97.

Hitchcock, P. F., Lindsey Myhr, K. J., Easter, S. S., Jr., Mangione-Smith, R. & Jones, D. D. (1992). Local regeneration in the retina of the goldfish. *J Neurobiol 23*, 187-203.

Hu, M. & Easter, S. S. (1999). Retinal neurogenesis: the formation of the initial central patch of postmitotic cells. *Dev Biol 207*, 309-321.

Jorstad, N. L., Wilken, M. S., Grimes, W. N., Wohl, S. G., VandenBosch, L. S., Yoshimatsu, T., Wong, R. O., Rieke, F. & Reh, T. A. (2017). Stimulation of functional neuronal regeneration from Müller glia in adult mice. *Nature 548*, 103-107.

Julian, D., Ennis, K. & Korenbrot, J. I. (1998). Birth and fate of proliferative cells in the inner nuclear layer of the mature fish retina. *J Comp Neurol 394*, 271-282.

Kassen, S. C., Ramanan, V., Montgomery, J. E., C, T. B., Liu, C. G., Vihtelic, T. S. & Hyde, D. R. (2007). Time course analysis of gene expression during light-induced photoreceptor cell death and regeneration in albino zebrafish. *Dev Neurobiol 67*, 1009-1031.

Kassen, S. C., Thummel, R., Campochiaro, L. A., Harding, M. J., Bennett, N. A. & Hyde, D. R. (2009). CNTF induces photoreceptor neuroprotection and Müller glial cell proliferation through two different signaling pathways in the adult zebrafish retina. *Exp Eye Res 88*, 1051-1064.

Kitambi, S. S. & Malicki, J. J. (2008). Spatiotemporal features of neurogenesis in the retina of medaka, *Oryzias latipes*. *Dev Dyn 237*, 3870-3881.

La Torre, A., Georgi, S. & Reh, T. A. (2013). Conserved microRNA pathway regulates developmental timing of retinal neurogenesis. *Proc Natl Acad Sci USA 110*, E2362-2370.

Lahne, M., Li, J., Marton, R. M. & Hyde, D. R. (2015). Actin-Cytoskeleton- and Rock-Mediated INM Are Required for Photoreceptor Regeneration in the Adult Zebrafish Retina. *J Neurosci 35*, 15612-15634.

Lenkowski, J. R. & Raymond, P. A. (2014). Müller glia: Stem cells for generation and regeneration of retinal neurons in teleost fish. *Prog Retin Eye Res 40*, 94-123.

Lust, K., Sinn, R., Perez Saturnino, A., Centanin, L. & Wittbrodt, J. (2016). *De novo* neurogenesis by targeted expression of atoh7 to Müller glia cells. *Development 143*, 1874-1883.

Madelaine, R. & Mourrain, P. (2017). Endogenous retinal neural stem cell reprogramming for neuronal regeneration. *Neural Regen Res 12*, 1765-1767.

Moshiri, A., Close, J. & Reh, T. A. (2004). Retinal stem cells and regeneration. *Int J Dev Biol 48*, 1003-1014.

Nagashima, M., Barthel, L. K. & Raymond, P. A. (2013). A self-renewing division of zebrafish Müller glial cells generates neuronal progenitors that require N-cadherin to regenerate retinal neurons. *Development 140*, 4510-4521.

Nelson, C. M., Ackerman, K. M., O'Hayer, P., Bailey, T. J., Gorsuch, R. A. & Hyde, D. R. (2013). Tumor necrosis factor-alpha is produced by dying retinal neurons and is required for Müller glia proliferation during zebrafish retinal regeneration. *J Neurosci 33*, 6524-6539.

Otteson, D. C., D'Costa, A. R. & Hitchcock, P. F. (2001). Putative stem cells and the lineage of rod photoreceptors in the mature retina of the goldfish. *Dev Biol 232*, 62-76.

Pavón-Muñoz, T., Bejarano-Escobar, R., Blasco, M., Martín-Partido, G. & Francisco-Morcillo, J. (2016). Retinal development in the gilthead seabream *Sparus aurata*. *Journal of Fish Biology 88*, 492-507.

Pollak, J., Wilken, M. S., Ueki, Y., Cox, K. E., Sullivan, J. M., Taylor, R. J., Levine, E. M. & Reh, T. A. (2013). ASCL1 reprograms mouse Muller glia into neurogenic retinal progenitors. *Development 140*, 2619-2631.

Qin, Z., Barthel, L. K. & Raymond, P. A. (2009). Genetic evidence for shared mechanisms of epimorphic regeneration in zebrafish. *Proc Natl Acad Sci USA 106*, 9310-9315.

Ramachandran, R., Reifler, A., Parent, J. M. & Goldman, D. (2010). Conditional gene expression and lineage tracing of tuba1a expressing cells during zebrafish development and retina regeneration. *J Comp Neurol 518*, 4196-4212.

Ramachandran, R., Zhao, X. F. & Goldman, D. (2011). Ascl1a/Dkk/beta-catenin signaling pathway is necessary and glycogen synthase kinase-3beta inhibition is sufficient for zebrafish retina regeneration. *Proc Natl Acad Sci USA 108*, 15858-15863.

Raymond, P. A., Barthel, L. K., Bernardos, R. L. & Perkowski, J. J. (2006). Molecular characterization of retinal stem cells and their niches in adult zebrafish. *BMC Dev Biol 6*, 36.

Raymond, P. A. & Rivlin, P. K. (1987). Germinal cells in the goldfish retina that produce rod photoreceptors. *Dev Biol 122*, 120-138.

Reichenbach, A. & Bringmann, A. (2013). New functions of Müller cells. *Glia 61*, 651-678.

Sharma, S. C. & Ungar, F. (1980). Histogenesis of the goldfish retina. *J Comp Neurol 191*, 373-382.

Sherpa, T., Fimbel, S. M., Mallory, D. E., Maaswinkel, H., Spritzer, S. D., Sand, J. A., Li, L., Hyde, D. R. & Stenkamp, D. L. (2008). Ganglion cell regeneration following whole-retina destruction in zebrafish. *Dev Neurobiol 68*, 166-181.

Sifuentes, C. J., Kim, J. W., Swaroop, A. & Raymond, P. A. (2016). Rapid, Dynamic Activation of Muller Glial Stem Cell Responses in Zebrafish. *Invest Ophthalmol Vis Sci 57*, 5148-5160.

Stenkamp, D. L. (2011). The rod photoreceptor lineage of teleost fish. *Prog Retin Eye Res 30*, 395-404.

Stenkamp, D. L., Powers, M. K., Carney, L. H. & Cameron, D. A. (2001). Evidence for two distinct mechanisms of neurogenesis and cellular pattern formation in regenerated goldfish retinas. *J Comp Neurol 431*, 363-381.

Thummel, R., Enright, J. M., Kassen, S. C., Montgomery, J. E., Bailey, T. J. & Hyde, D. R. (2010). Pax6a and Pax6b are required at different points in neuronal progenitor cell proliferation during zebrafish photoreceptor regeneration. *Exp Eye Res 90*, 572-582.

Thummel, R., Kassen, S. C., Enright, J. M., Nelson, C. M., Montgomery, J. E. & Hyde, D. R. (2008). Characterization of Müller glia and neuronal progenitors during adult zebrafish retinal regeneration. *Exp Eye Res 87*, 433-444.

Ueki, Y., Wilken, M. S., Cox, K. E., Chipman, L., Jorstad, N., Sternhagen, K., Simic, M., Ullom, K., Nakafuku, M. & Reh, T. A. (2015). Transgenic expression of the proneural transcription factor Ascl1 in Müller glia stimulates retinal regeneration in young mice. *Proc Natl Acad Sci USA 112*, 13717-13722.

Vecino, E., Rodríguez, F. D., Ruzafa, N., Pereiro, X. & Sharma, S. C. (2016). Glia-neuron interactions in the mammalian retina. *Prog Retin Eye Res 51*, 1-40.

Vihtelic, T. S. & Hyde, D. R. (2000). Light-induced rod and cone cell death and regeneration in the adult albino zebrafish (*Danio rerio*) retina. *J Neurobiol 44*, 289-307.

Wan, J., Ramachandran, R. & Goldman, D. (2012). HB-EGF is necessary and sufficient for Muller glia dedifferentiation and retina regeneration. *Dev Cell 22*, 334-347.

Wan, J., Zhao, X. F., Vojtek, A. & Goldman, D. (2014). Retinal injury, growth factors, and cytokines converge on beta-catenin and pStat3 signaling to stimulate retina regeneration. *Cell Rep 9*, 285-297.

Wan, Y., Almeida, A. D., Rulands, S., Chalour, N., Muresan, L., Wu, Y., Simons, B. D., He, J. & Harris, W. A. (2016). The ciliary marginal zone of the zebrafish retina: clonal and time-lapse analysis of a continuously growing tissue. *Development 143*, 1099-1107.

White, D. T., Sengupta, S., Saxena, M. T., Xu, Q., Hanes, J., Ding, D., Ji, H. & Mumm, J. S. (2017). Immunomodulation-accelerated neuronal regeneration following selective rod photoreceptor cell ablation in the zebrafish retina. *Proc Natl Acad Sci USA 114*, E3719-E3728.

Wilson, S. G., Wen, W., Pillai-Kastoori, L. & Morris, A. C. (2016). Tracking the fate of her4 expressing cells in the regenerating retina using her4:Kaede zebrafish. *Exp Eye Res 145*, 75-87.

Xiang, M. (2013). Intrinsic control of mammalian retinogenesis. *Cell Mol Life Sci 70*, 2519-2532.

INDEX

C

D

H

I

L

M

N

R

S

T

U

V